A Case Study in
Archaeology

A Student's Perspective

Second Edition

Pambamarca Archaeology Project (PAP)

Mary C. Sullivan
Samuel Connell

Kendall Hunt
publishing company

Cover images provided by Mary C. Sullivan and Samuel Connell.

Kendall Hunt
publishing company

www.kendallhunt.com
Send all inquiries to:
4050 Westmark Drive
Dubuque, IA 52004-1840

Copyright © 2009, 2012 by Kendall Hunt Publishing Company

ISBN 978-0-7575-9776-3

Printed in the United States of America
10 9 8 7 6 5 4 3 2

Contents

Acknowledgments

We owe our gratitude to a long and diverse list of people. To start with, we must acknowledge the major force behind the creation of this research program, Don Carlos Perez. Our research program was conceived by Don Carlos decades before our arrival on his doorstep. Imagine the joy he felt having archaeologists drop into his lap as he began selling us on the idea of a project at Pambamarca. And imagine our joy at being so welcomed. In reality, Don Carlos gets all the credit for defining the central question of our investigations about the identity and origin of the people who lived in the Pambamarca fortresses. He was a gentleman and a historian. He is sorely missed.

There are many more people to whom we are indebted. In Quito, thanks must go to the wonderful team of professionals at the Instituto Nacional del Patrimonio Cultural, and the Departamento de Arqueología led by Mónica Bolaños, Francisco Sanchez, Maria Moreira, and Rosillo Murillo.

In Cangahua, we are grateful to Padre Roberto Neppas, who has provided his warm friendship and collaboration since 2001. We eat our meals, sleep in our sleeping bags, and conduct our meetings in the communal house connected to his church. Elena Tandamayo, Elena Gualavasi, Maria Gualavasi, David and Mario Quimbulco, and Franklin Imbaquingo take great care of us, helping our project members stay healthy and happy with excellent meals. In addition, there are many others who provide unconditional help in the town of Cangahua, including those who clean, transport, bake bread, and wash clothes.

Also, we extend a very warm thank you to the Hacienda Guachalá and Don Diego Bonifaz, Gabriel Bonifaz, and Cristóbal Cobo, who help us with friendly enthusiasm and constant intelligent advice.

Thanks also to the members of the communities of Pambamarca, especially those of the community of Chumillos Central and señor Melchor Farinango who has helped us incredibly since the year we began research in 2002.

Our research in Ecuador follows in the giant footsteps of important and helpful scholars; many of them have provided suggestions and are always welcome to visit. These include Ernesto Salazar, David Brown, Florencio Delgado, Tamara Bray, Karen Stothert, Mariuxi Cordero, Manuel Coloma, Stephen Athens, Pepe Echeverría, Patty Mothes, Peter Hall, Ronald Lippi, Marc Becker, Larry Conyers, Brandon Lewis, and Alejandra Gudiño. We also owe a large debt of gratitude to all the Ecuadorian students that have contributed to the project over the years, including Julio Mena, Oscar Cajas, Carlos Montalvo, y muchos más.

We also are thankful for the contributions of our home institutions and their representatives, including Foothill College, UCLA, and Columbia University. And many thanks to all the students who worked during the field seasons, not only for your hard work but your important research contributions. To our staff, without whom we would have never been able to succeed in the work, thank you so very much for all your help. Without you the work at PAP would never have come close to being completed. Thanks for teaching, helping, and listening.

Finally, we send a special thanks to the Gonzalez family for their love and support. Especially Fanny. Also, we thank Bill Allen for his support during the writing of this book and Andrea Sullivan for her helpful editorial remarks. Bryan Haley kindly reprocessed all of the GPR data, aided in the improvement of figures, and provided invaluable advice. Mark DeGuzman is heartily appreciated for improving many graphics. We also thank Kendall Hunt for support.

Preface for Instructors

A Case Study in Archaeology: A Student's Perspective is a workbook intended to supplement the main textbook/coursework in an entry level anthropology or archaeology class. *A Case Study* is interactive: there are opportunities to think about, criticize, react to, and engage the material presented. Here a student will go beyond the theories of archaeology and instead delve into practice. Critical thinking, discussion, group work, and learning to express oneself verbally and in writing are crucial skills needed not just in anthropology but in any college course. This workbook is intended to be a springboard for such activities. Many of the questions and exercises could be conducted individually, in small groups, or class-wide. Because it's so flexible, the book can be modified to account for time constraints and individual teaching preferences and goals.

The data in this workbook are from the Pambamarca Archaeology Project (PAP) located high in the Andes Mountains of northern Ecuador. PAP investigates the intersection of pre-Inca, Inca, and Spanish cultures. Neither the instructor nor the student needs to be familiar with the project area because we take care to introduce the culture history, background, findings, methodologies, and communities surrounding our project. From there, the goal of each exercise is to reinforce the reading, think about how our contemporary world mirrors the past, relate the activities of these groups to a student's own life, consider ethical dilemmas, and interpret data. Some students may initially feel intimidated by the open-ended nature of the questions. In fact, the "right" answer is one that is thoughtfully and thoroughly considered and expressed. The goal of these exercises is to improve a student's critical thinking, deductive reasoning, writing, and communication skills.

We have used real data from PAP precisely because their ambiguous nature recreate the challenges of real life, such as having incomplete datasets, struggling with ethical issues, interpreting features during excavation, trying to survey a vast area, and communicating with the local population. Because the examples from this book are from an ongoing investigation, students experience what the field is like: having to make preliminary analyses even though the data are sometimes unclear. The answers to the questions in *A Case Study* are not preconceived, but they are intended to be found using a combination of data available from the text and general deduction. The instructor can use the opportunity to show students how to use all visual clues and background knowledge to make educated hypotheses. The professor could also offer alternative views on the way PAP conducts its project. The open nature of the exercises should lead to vigorous group and class discussion.

A Case Study covers what we consider to be the main facets of field school: learning the basics about the discipline of archaeology, culture history of a site, and elementary field methodology. It also provides an introduction to applied anthropology. The first chapter starts at the beginning: how do you know what to study? Where to dig? What questions should you ask? We explain how archaeologists try to reconstruct the lives of historic people by finding their things, and we encourage the students to consider how their own lives are represented by their belongings. Finally, Chapter 1 reveals a brief cultural history of PAP. We discuss local resistance to imperial invaders, colonialism, and the process of social change that occurs when cultures compete for dominance. These issues are prominent at a project like PAP's where three rival culture groups overlap.

In Chapter 2, we talk about field methods such as survey, excavation, and laboratory analysis. PAP has had a geophysics program since its early days, and we specifically discuss ground penetrating radar (GPR) to identify sites. We think it's an advantage to expose students to such methods as more archaeological projects now utilize remote sensing. Even novices can practice identifying visual spatial relationships to find cultural remains. We also cover excavation, stratigraphy, and recording data, and we challenge students to identify features found during a dig. Finally, we end in the lab where we talk about artifacts and ask students to identify examples.

Our final chapter is about applied anthropology and ethics. PAP is based in a small town north of Ecuador's capital city of Quito. This area is home to many indigenous populations, mestizo groups, hacienda owners, and other potential stakeholders. Our project's presence has an enormous impact on the local population because of the influx of money and foreigners. Some of the surrounding areas didn't have electricity when the project began in 2000, and they certainly had very few visitors. We want students to consider the role of the archaeologist in the local community, the ethical obligation to consider one's footprint, and how to balance the needs of science and the local populace.

Finally, A Case Study is written from "A Student's Perspective," which means there is commentary and feedback from real students throughout the book. We try to recreate the experience of attending field school for the first time. There are plenty of "how-to" books on the market for archaeology, but many of them engage higher level students. This is meant for the novice; it's intended to interest the reader, be easy to read, and be interactive. Just as there are many interpretations of artifacts and architecture in the field, we wanted students using this book to deliberate the issues amongst themselves, discuss the data, and come up with their own conclusions. It's a great chance for them to practice the intellectual collaboration and critical dialogue that is the basis of scholarship. And, we hope they have fun trying!

The second edition of this book has improved figures, clarified assignments, and additional instructions. To aid instructors who use the book, we have provided suggestions on how to modify the exercises to suit the individual needs of the class in which it's taught. Look for the "Instructor's Note" icon throughout the text for suggestions on how to adapt the activities for homework or class discussion. We've also included ideas about how to expand on the learning outcomes for each chapter and turn them into more in-depth research projects. We hope A Case Study provides enough flexibility for it to be an enjoyable, useful, debate-inducing tool to supplement any entry level anthropology class.

Introduction to the Text

Did you ever wish you could traipse through the jungle with Indiana Jones to dig up ancient artifacts? Do you find yourself spending countless hours in front of the History Channel or the Discovery Channel? If you've ever wanted to know what goes on at those digs, wondered what an archaeologist does, or considered volunteering at an archaeology project, then *A Case Study in Archaeology: A Student's Perspective* is here to help! This is a workbook that will teach you about methodologies, lifestyle, ethics issues, theory, and the real "dirt" on an archaeology project.

A Case Study is unique because this book is a two-way street. Every time we share new information about our field, you'll be asked to respond in a series of thought-provoking, controversial, conversation-sparking questions. You'll have a chance to think for yourself, to try it on your own, to take the knowledge you've gained and apply it to real-life archaeological problems. Our goal for you is to learn about the culture history of our site in Ecuador, understand basic field methodology, and attempt to navigate the ethical issues of fieldwork. Oh yes, and we want you to have fun doing it!

We're sharing the details of our project, the Pambamarca Archaeology Project (PAP) simply because we know it best, and therefore, it makes a great case study. There are many projects around the world with varying philosophies, theory, methodologies, and approaches to the field. We're simply one example of the way it's done. We're willing to expose the complications, successes, missteps, and questions of PAP in order to spark healthy debate about the practice of archaeology. In real life, the answers are not always neat, and the solutions are rarely perfect. By using an actual, ongoing project as an example, we can recreate the invigorating experience of being in the field.

Chapter 1

Becoming an Archaeologist and the Culture History of Pambamarca

This chapter starts at the beginning: how do you become an archaeologist and how do you know where to dig? We'll talk about what archaeologists are interested in, what types of questions they want to answer, and what we can learn about ourselves by studying the past. We'll also discuss what archaeologists are looking for in the field and how artifacts tell the story of cultures that have come and gone. Finally, we'll reveal the history of our project area and the culture groups we're studying.

So, why the Pambamarca Archaeology Project (PAP), and what makes it so special? To begin, it's located in an amazing landscape. It's in the country of Ecuador, which is on the northwest side of South America (Figure 1.1). The Andes mountains run right through the middle of the country, with the Pacific coast to the west (along with the Galapagos Islands) and the Amazon-region to the east. This small country has more biodiversity than most other countries in the world. But the part we're interested in is a small community, high in the Andes, on the equator, north of Ecuador's capital city of Quito (Figure 1.2).

The Pambamarca project area is impressive looking: it contains ancient Inca and pre-Inca fortresses built on top of 12–14,000 foot tall mountains. A Spanish hacienda built in 1580 that looks like it was transported from the set of Zorro also sits on the project. In some ways, visiting this place is like walking

FIGURE 1.1 Location of Ecuador in South America. © Martine Oger, 2009. Shutterstock, Inc.

back in time. You can almost imagine the Inca in their tunics hurling sling stones from their fortress peaks or see Spanish conquistadors riding horseback through the hacienda's courtyard. Our project is attempting to understand how the pre-Inca, Inca, and Spanish were working, fighting, living, partying, and surviving in this chilly, craggy equatorial zone. PAP is the organization investigating such questions.

FIGURE 1.2   Note the location of Pambamarca on a topographical map of
Ecuador. © Olinchuk, 2009. Shutterstock, Inc.

GETTING STARTED

Where and how to start an archaeological project depends on a lot of factors. Maybe the
project is in a culture area you find interesting—perhaps you've always had a "thing" for
South America or Egypt or the Maya. Or maybe the project is using a field method you want
to try, like geophysics, aerial photography, or satellite images to find sites. There are many
reasons to join an archaeological project. To be honest, it's part academics, part logistics
(money, professors, backing), part research question, and sometimes, just a little bit of fate.

In order to explain how an archaeologist knows where to dig, what questions to ask,
and how to approach this subject, let's look at the true story of how PAP was born. Here
Dr. Samuel Connell, director of PAP, explains how he and his collaborators got started.

"My wife's (Ana) family is from Ecuador, so we
were down visiting for a couple of months when
we decided to go see some archaeological sites.
We weren't really sure where to go, but we had a
touristy book about Ecuador that included
a vague paragraph about a place called
Quitoloma. It was supposed to be a massive Inca
structure, and it sounded impressive so we figured, why not go there?

We borrowed Ana's Dad's Isuzu Trooper and headed out to go find the place. As far as we could tell, Quitoloma was supposed to be north of Quito (the capital of Ecuador) and high in the Andes, off these unmarked roads. Needless to say, the first day we get lost. But we try again the second day. And that's when we rolled into the small, windswept village of Cangahua. Many of the people who lived there were indigenous and some still spoke Quichua, the language of the Incas.

Sitting on the street corner were a number of older men, so we asked them how to get to Quitoloma. They said, 'Oh. Quitoloma. First, you must speak to Don Carlos Perez.' We had no idea who this person was, but we shrugged and asked directions to his house. Maybe he knew something about the site. After considerable pounding on a modest front door, Don Carlos arrived—short, leathery, and gray. We were shown into an antechamber in the house where he received guests, and then we were left there. Alone. For some time.

It was the classic anthropologist moment when you are wondering what the heck you are doing, and you have placed yourself in an awkward situation. But yet something inside of you says, 'follow this through, things will work out, they always have.' People are people wherever you go, and they never fail to surprise!

Finally, Don Carlos returned to the room. He teetered in, sat down, observed us for a moment, and finally said, 'Ah. So. You want to see . . . Quitoloma?' We explain that we're simply seeking directions, but he tells us that it is too late in the day to go and he is too sick to show us the way. We told him that as archaeologists, we simply want to see the site and we don't expect him to be our guide.

At the word 'archaeologist,' Don Carlos' manner changed. A gleam appeared in his eye. He looks at us directly and said, 'You. You. Will. Dig. Here.' I raised my eyebrow at my wife. This was too much like a movie. He started to tell us about a war, the Incas, the resistance, the fortresses, and the Spanish. Eventually, Don Carlos continued, 'You will come back here. You will come back in two weeks. Then, I will take you to . . . Quitoloma.'"

"The PAP field school was my first fieldwork experience, but I was already enrolled in a graduate program. I had never been to South America or the Andes before and knew very little about Andean or Incan archaeology, except what I learned out of a textbook. The first day hiking to Quitoloma was an inspiring and amazing experience I will never forget."

—Matt Schauer, 27, of the University of Illinois at Chicago, Ph.D. student

"I have always been really interested in the ancient parts of the past. I loved looking at National Geographic magazines, loved museums, and loved taking world history courses in high school. As an undergrad, I found myself naturally drawn to the course descriptions for archaeology. By the end of my sophomore year, I had enough credits to be an anthro major."

—Amber Kling, 29, of State University of New York at Buffalo, Ph.D. student

How could they say no? Ana and Sam agreed to return, and when they did, Don Carlos took them up ancient Inca roads, through fields and pasture, until the road ran out and they drove on grass. Then all three of them climbed to 12,000 feet above sea level to view Quitoloma for the first time. Ana and Sam took in the mammoth structure, glanced at each other, and knew Don Carlos was right. They would dig. Right there.

Exercise 1.1: Think about an area in your hometown or a place you visited while traveling that interested you. Where would you start an archaeological project? What group of people or places do you want to know more about? Why?

_____

_____

_____

_____

Exercise 1.2: Is there a question you've always had about the past, about the nature of humans, or why people do what they do? Write a research question that you want to ask.

_____

_____

_____

Introducing the Pambamarca Archaeology Project (PAP)

What Don Carlos showed Sam and Ana was a gigantic ancient Inca fortress built on top of a tall peak, high in the Andes. Quitoloma is made of concentric circles of stone—multiple walls to keep invaders out. If you see Quitoloma from far away, it looks like a crown on top of the mountain (Figure 1.3). The fortress is over 300 yards wide by more than 500 yards long. That is three football fields wide by almost five football fields long. When you stand on top of Quitoloma, you're so high up that you can see Ecuador's capital city of Quito—over twenty miles away. As Sam and Ana observed the landscape, they noticed other fortresses ringing nearby peaks.

The fortresses, including Quitoloma, in this remote vertical landscape are in an area called Pambamarca. PAP was named after this region. Even today, Pambamarca is a

FIGURE 1.3 Quitoloma's concentric circles with someone standing at its top. Also see additional fortresses near the horizon. Courtesy of PAP/Samuel Connell.

remote place. No one lives in such a high elevation zone (called the páramo) because no crops can grow there. Quitoloma itself is surrounded by tall blonde grass, bleached by the sun searing it in a thin atmosphere. There's nothing else around except the other fortresses on the surrounding hilltops. This is where the wheels in Sam and Ana's brains started to turn. Why would the Inca bother building a grand fortress in such an isolated area? After all, the Inca capital, Cuzco, was over a thousand miles away in Peru.

According to historical accounts, the Inca put serious effort into fighting for this remote outpost located precisely on the equator. The Inca Empire was larger than Rome and very powerful, but despite its strength, the native Ecuadorians proved to be formidable opponents. They fought long and hard for their liberty, and they succeeded in resisting the Inca for as many as 20 years—far longer than most other groups in western South America.[1] The Pambamarca landscape represents the collision of manifest destiny and a battle for freedom.

Sam and Ana were flooded with potential research questions. How did this Andean David manage to hold off the Inca Goliath? Why did the Inca want to own this land—seemingly in the middle of nowhere? Who were the indigenous Ecuadorians? How did the Ecuadorians' lives change when the Inca came? Everyone knows that the Inca no longer rule South America; they were overtaken by the Spanish. What was different after Pizzaro and the Spanish arrived?

Sam and Ana decided that they had to find out how these three culture groups (the pre-Incans, the Incans, and the Spanish) lived in this area. Therefore, they started the Pambamarca Archaeological Project, PAP, to do just that. The first field season took place in the year 2000. It was a tiny operation—the two of them and a couple other friends they'd worked with in the past. They added Sam's former colleague from Maya archaeology, Chad Gifford, as a co-director. Since then, the project has grown to include more than 70 archaeology students per field season and plenty of staff members.

Even though Sam, Ana, and friends have been working in the area for over ten years, there are many questions left to be answered about the culture groups in this area. For instance, why did the Cayambe♦ build fortresses *before* the Inca invasion (Figure 1.4)? What

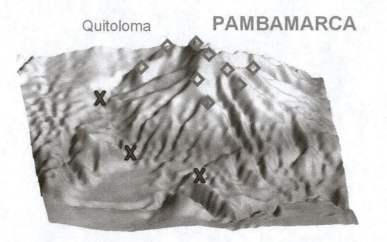

FIGURE 1.4   Pambamarca fortresses on the landscape. Diamonds are Inca fortresses, and Xs are indigenous (Cayambe) fortresses. Courtesy of PAP/Samuel Connell.

♦ "Cayambe" is the name PAP is giving to the indigenous Ecuadorian population within the PAP borders. There were many distinct indigenous culture groups in the area. They are sometimes called the Pais Caranqui (who were known as mound builders), which is a conglomerate of all groups found south of the El Chota River and north of the Guayllabamba River. For the purposes of this book, we'll use the term "Cayambe," but note that when we refer to "mound builders," we mean the Cayambe and Pais Caranqui groups as well.

was the Cayambe political structure like prior to the Inca? After all, the mound-building Cayambe had been farming and living in the area for over 4,000 years before the Inca showed up.[2] They had a rich social life, political structure, and economy before foreigners came. What was that life like? What was important to them?

And there were other issues to be addressed. Soon after the Inca conquered the Cayambe (perhaps as little as ten years later), the Spanish arrived, and we all know who won that battle. The Spanish colonial settlement most notable within PAP borders is the Hacienda Guachalá, established nearly 500 years ago. What was it like to be colonized twice? What lasting ramifications are still felt in Pambamarca from the intersection of these groups today? Is there tension between the indigenous, the mixed people, and the hacienda owners? How could these tensions impact the ability of archaeologists to conduct their work?

Questions like these are what archaeologists really love to sink their teeth into. An archaeologist's primary goal is to understand people. For instance, learning about Pambamarca can speak to larger issues about dominance and resistance, borderlands and frontiers, conquest and subjugation, and social transformations and integration.[3] What is it like to have your land conquered? What is it like to go from being free to being under control of an invading force? How do small groups resist much bigger ones? What changes occur within a group when it is taken over?

It's common for history to be written from the point of view of the winner. After all, how much do you learn about the Native Americans in an American history class? However, PAP thinks the point of view of local group is just as important as that of the colonizer.[4] It's not like colonization is a one-way street: the reaction of a local population has consequences for imperial expansion and colonial aspirations.[5] The 'Goliaths' of the world might be big and powerful, but they cannot assume everyone will bow to their wishes. The way in which the underdog responds to provocation is interesting, and could have implications for contemporary society.

INSTRUCTOR'S NOTE: These exercises can be done individually or as a springboard for class discussion.

Exercise 1.3: Think about the contemporary world. Are there any superpowers extending their reach into other people's countries? If so, what is the local reaction? Is there ever a time in today's day and age where a country has more money, more military, or more technology—yet still struggles to defeat the locals? Explain why seeming superiority doesn't always translate into immediate success.

_____

_____

_____

_____

_____

_____

_____

Exercise 1.4: Why do groups go to war against each other in modern times? What were historical motivations for war?

_____

_____

_____

_____

Exercise 1.5: Why do you think some cultural groups or countries don't get written about (much) in history books? Why is history usually written from the perspective of the dominant group? For instance, in standard American history courses, limited attention is given to Native Americans compared to European Americans. Why do you think this is?

_____

_____

_____

_____

_____

Exercise 1.6: Why is it beneficial to remember and understand people who were conquered? What lessons can we learn from the "losers" of history?

_____

_____

_____

_____

Exercise 1.7: Is contemporary Ecuador a major player on the world stage? Why or why not? What would it take for Ecuador to change its status on the world "map" today?

_____

_____

_____

"Everyone wants to believe that Ecuador became civilized when the Inca conquered the northern native tribes. But it is clear that there were many groups already living in this region that were involved in advanced systems of exchange. They had a highly organized work force with the ability to build ramped truncated pyramids and make beautiful pottery and textiles."

—Ana Gonzalez, University of Hawaii, co-director of PAP

"Pambamarca is an ideal place to study a fundamental and fascinating question in anthropology: what roles do local history and culture play within broader political and economic structures during instances of culture contact under imperial and colonial expansion? For me, any information about colonial and imperial encounters between the Incas and indigenous Ecuadorians is what it is all about!"

—Chad Gifford, advising dean of Columbia University, co-director of PAP

UNDERSTANDING MATERIAL CULTURE (ARTIFACTS)

Earlier in the chapter, we started talking about how archeologists are interested in stories of people from the past. For instance, we want to know what life was like for the Cayambe, and how their lives changed with the arrival of the Inca and Spanish. Other archaeologists might be interested in these people's religion, family life, or the relationship between men and women. Understanding the intricacies of an entire people is a big undertaking. There are countless questions to be asked and answered.

Of course, we are talking about events that happened hundreds and even thousands of years ago. Everyone who participated in these cultures is dead by now. Unless we get a Ouija board, we cannot ask ancient people questions. Furthermore, neither the Cayambe nor the Inca wrote anything down, so we can't read about it either. How do we gain access to this information? The answer is in the artifacts. Artifact is another word for material culture, and material culture is the stuff people use in their daily lives.

If archaeologists are detectives, then artifacts are our clues. By finding *things* people used in the past, we can make inferences about what those people did, felt, or believed in. Imagine there are three students in your anthropology class who allow us to examine the contents of their backpacks. One student has a Bible. Another has tennis shoes, a water bottle, and shorts. The last student has an apron, black pants, a white shirt, and a tie. Without asking these students any questions, can you guess what each person might be doing after class? Or what is important to them? This is what archaeologists do with artifacts.

The things you do in your daily life have a footprint. These physical clues—artifacts—are what archaeologists use to reconstruct ancient people's lives. With artifacts, archaeologists can figure out what people did with their day, where they went, who they met, what was important to them, how their family and social life was structured, what their religious or spiritual beliefs might have been, and how they dealt with stressors. Being an archaeologist is a lot like being a history detective! Working backwards from clue to activity isn't always easy, but the following exercises will help.

"Having never worked on a project before, living in the area where we were working was new to me. I had only ever seen artifacts in a lab, disconnected from their origins. I had not fully realized the practical implications of the level of documentation that is required: developing a recording system, a system for cataloguing artifacts, etc. These are things that are very well and good to read about, but until I saw the system in action, the importance of having all of these elements planned out in advance had not struck me."

—Perri Gerard-Little, 22, recent grad from Columbia University

Exercise 1.8: Make a list of five activities you *do* often. They need not be part of your daily routine, just something you do on a regular basis. For example, if you do something semi-annually, you could write down that activity. Also, be sure to think of distinct activities. Eating breakfast, lunch, and dinner can be collapsed into one activity: eating.

_____

_____

_____

_____

_____

Exercise 1.9: Now think of at least three items—physical things—that you use to facilitate each of the tasks listed above. For instance, a gardener might write that he/she needs gloves, a shovel, fertilizer, and a water hose. If you can think of more than just three items per task, list those too. When you're done, you should have a list of at least 15 things on the paper.

_____

_____

_____

_____

_____

_____

_____

Exercise 1.10: Now think of the places that you go to in order to complete these tasks. Try to think of everywhere you can. If you work out every day, then you go the parking garage to get your car, you use streets and highways to reach the gym, and then you go to the gym itself. Those are three places—at least—that exist just so you can exercise.

_____

_____

_____

_____

_____

Exercise 1.11: Let's practice thinking like an archaeologist. Pretend the following lists are the artifacts found during an archaeologist's dig. Determine what activity they represent. You might come up with more than one activity for each group of artifacts.

Scenario 1: Sneakers, MP3 player, towel, water bottle. What type of activity might these represent? Where are these things most likely to be found?

_____

_____

_____

Scenario 2: Plate, fork, chicken bones, wine bottle. What type of activity might these represent? Where are these things most likely to be found?

_____

_____

_____

Scenario 3: Book, pencil, chalk, desk. What type of activity might these represent? Where are these things most likely to be found?

_____

_____

_____

Exercise 1.12: Pretty easy, right? When you're in the field, sometimes things aren't so clear-cut because the artifacts you uncover will not necessarily be used in your contemporary life. However, you're on the right path toward thinking like an archaeologist. This next part is a bit harder. Now consider the person who used these artifacts. Where and for what purpose were these items used?

Scenario 1: What values does a person have who uses these items?

_____

_____

If there is a special place set aside for this activity? What does it say about the daily lives of these people?

_____

_____

_____

_____

Scenario 2: In what country/ethnic group/culture might you find the items in this scenario? Why?

_____

_____

_____

There are a few groups who are unlikely to use these items. Who are they and why would they not use them?

_____

_____

_____

Scenario 3: Can you tell how old the people are who use these items? What are they doing? How do you know? What other artifacts might help you determine age?

_____

_____

_____

How does this scenario reflect a society's values?

_____

_____

_____

 INSTRUCTOR'S NOTE: In order to do the "Group Fun Option," students must complete Exercises 1.8–1.10.

GROUP FUN OPTION:

INSTRUCTOR'S NOTE: Once the students have completed the previous exercises, break them into small groups or have them choose a partner. Then have them work through Challenges 1–3. Note that Challenges 4 and 5 might best be conducted in small groups or as a class-wide discussion. The responses can be written on the chalk/white board for the class to see.

Challenge 1: Read the physical things (artifacts) that you use to complete your tasks (from Exercise 1.9) out loud. See if your partner can guess what tasks you have to do each day. How many did your partner get right? Record your scores here.

_____

_____

_____

Did you or your partner get any wrong? What stumped you and why?

_____

_____

_____

_____

Challenge 2: Now get a new partner. Let's challenge this partner. Look at the list of artifacts you have for each task. Now scramble them and list them on a separate piece of paper so they are jumbled together. Now exchange lists. Can your partner find the artifacts that go together now that they're mixed up? Can your partner figure out what activities you're representing? Report how you each did below. Were there any that stumped you? What caused the confusion? If you got most of them right, what made it easy?

_____

_____

_____

_____

_____

_____

Challenge 3: Find a third partner. Look for someone who seems like they're up for a challenge! This time, look at the list you made for your Challenge 2 partner. This is the big jumbled list of all of your daily artifacts. Remove any item that would disintegrate after 500 years in the ground. Rewrite your list. Give it to a new partner. Can he/she pick any artifacts that go together? Can he/she guess any of your daily activities from this list? Record your success and fail rates. Explain why you were successful (or not).

_____

_____

_____

_____

Challenge 4: Ultimate group challenge: Who. Have partners present their Challenge 3 lists to the class (this is the list of artifacts MINUS the artifacts that would deteriorate after 500 years). If the class correctly guesses which activity is represented by those artifacts, is there any way to determine which gender conducted the activity? How about the age of the participants? Discuss the challenges of this activity in class.

Challenge 5: Ultimate group challenge: Patterns. Archaeologists try to see if there are any patterns in the activities or artifacts of the people they're studying. If we find patterns in material culture, we might be able to guess what time period we're studying or what culture groups we're investigating. For instance, if we find that people play with more Xboxes than Rubik's Cubes, what does that tell us about the time period we're dealing with? Finally, are there any patterns that we can detect about the activities you and your classmates participate in? Based on your artifacts, do you have similar schedules or like the same things? List the top five artifacts and activities that occur with the most regularity. What do these things say about your values? Your environment? Your socioeconomic status? How can we tell?

THE CULTURE HISTORY OF PAMBAMARCA

However exciting it is to strap on your boots and literally head for the hills with your shovel and trowel, archaeological projects first begin in the library. In order to know what to expect or where to dig your first hole, you need to familiarize yourself with what is already known about the area. There are often historical resources available from other archaeologists who have worked in the vicinity prior to you. First learning what has already been discovered or known will guide your research and help you decide what you're most interested in. Part of the scholarly process involves learning from others instead of always starting from scratch. In this part of the chapter, we'll discover a little bit about the lives of the Cayambe, Inca, and Spanish in Pambamarca.

TABLE 1.1   Time Periods for the Cayambe People.

PERIOD	TIME
Pre-Ceramic	9000–1550 B.C.
La Chimba	700 B.C. – 1 A.D.
Early Intermediate	1–700 A.D.
Late Intermediate	700–1250 A.D.
Late Period	1250–1505/1525 A.D.
Inca Period	1505/1525–1534 A.D.

THE CAYAMBE

Although comparatively little archaeological research has been done on the Cayambe, Table 1.1 reflects known cultural changes through time for this group. Each time period represents a major technological or cultural advancement. For instance, the "La Chimba Period" represents the time when the Cayambe first started to make pottery. The earliest Cayambe sites within the Pambamarca region date to the Late Intermediate Period (the earlier artifacts were found by other archaeologists nearby). However, for the sake of this chapter, we'll skip ahead to the Late Period because we know the most about this time, and it's a good example of what life was like in the years before the Inca came—when the Cayambe were free.

The Cayambe Late Period

The Late Period is well-known because this is when the Cayambe (and other northern Ecuadorians like the Pais Caranqui) started building their famous large mounds, and these distinct structures are hard to ignore. They're usually quadrilateral in shape with sizeable ramps (as much as 1,000 feet long) on one side. They can be over 100 yards long and 65 feet tall (Figure 1.5). That means the Cayambe were making dirt mounds equivalent to a football field in length and the height of a six-story building. Now consider these were constructed with no bulldozers, horses, wheels for carts, or cranes. When you realize the magnitude of manual labor used to construct them, these enormous square-shaped piles of dirt take on a whole new significance.[6]

The Cayambe built their mounds out of a material called *cangahua*. (Not to be confused with the town of Cangahua.) *Cangahua* (Figure 1.6) is really hard volcanic ash that can be cut into blocks.[7] There are plenty of volcanoes in the region (after all, it is the Andes) so *cangahua* would have been plentiful. When we survey or excavate in Pambamarca, we generally expect to find Cayambe sites made out of *cangahua*. Understanding the typical building materials for a cultural group helps to narrow down who might have been using a given site. This is very important knowledge in a place like Pambamarca where you have three distinct cultural groups in the exact same area.

FIGURE 1.5   Quadrilateral mounds dotting landscape. Courtesy of PAP/Samuel Connell.

FIGURE 1.6   *Cangahua* blocks used as foundation for a house. Courtesy of PAP/Samuel Connell.

A question that has been plaguing archaeologists in northern Ecuador for a while is: what were these huge mounds used for? Some think they were the houses of very important local chiefs,[8] almost like a governor's mansion. Domestic artifacts like pottery and tools, as well as round floors of houses, have been found on top of the mounds. These artifacts are what led archaeologists to draw this conclusion. Other archaeologists think that the mounds contain too few domestic artifacts to indicate habitation. Instead, they think the mounds were the site of huge parties or community gatherings.

Exercise 1.13: These mounds are gigantic, and they took a lot of man-power to build. We find large, public structures and monuments in our every day lives. Name two and explain what they each represent about the priorities or values in our culture.

_____

_____

_____

_____

The magnitude of the mounds indicates one thing for sure: the Cayambes had a centralized government. It takes leadership to motivate people to haul that much *cangahua* into a big square platform. These mound groups are scattered equally over the northern Ecuadorian landscape, and no mound group is particularly larger than the next. Whichever chief convinced or coerced people to build the mounds likely ruled the area surrounding his mound. Therefore, there were probably many separate chiefdoms all through the area. However, just because these people all built mounds and lived in northern Ecuador doesn't mean they were friends.

Paz Ponce de Leon,[9] a Spanish explorer who visited the area in the 1500s, described the constant in-fighting among the Cayambes and their fellow indigenous Ecuadorians.

"[They] brought war upon one another about the land they possessed, and the one that was most able displaced the other of all he possessed; and these controversies the Indians have always had with their neighbors, of a manner that was all disorder."

© Bettmann/CORBIS.

The problems between the neighboring chiefdoms probably had to do with turf battles. Some archaeologists think they were trying to protect limited food resources from each other.[10] After all, it's not easy to grow food at over 10,000 feet above sea level. Other people believe the chiefs were guarding lucrative trade routes. Either way, despite the conflicts between individual mound-building chiefdoms, they all had one thing in common: they hated outsiders even more than each other. The Cayambe's natural distrust of foreigners could have been their best defense against the future Inca invaders.

Keep in mind that the Cayambe chiefdoms were fighting one another to protect scarce resources. It was bad enough that their neighbors sometimes tried to gain access to their crops or trade routes. The last thing they needed was some foreign group taking their land and hence, their way of making a living. A foreign group could also force them to change their culture, which is something that could affect all Cayambes. So, when push came to shove, the Cayambe chiefdoms would ban together to get rid of invaders because it benefited all of them. It's sort of like family. You might insult your brother or sister, but if someone else tries it, you'd protect your sibling.

Some scholars say[11] that people use the concept of ethnicity to symbolize their group and define themselves as different from other people. In defining themselves as ethnically exclusive, they are in opposition to other cultures or ethnicities. These ethnic identities keep the group insular: that means it keeps those who ethnically belong "in" and anyone who is different "out." These ideas can lead to competition with other groups that appear to be outside the ethnicity or culture. Maybe by practicing exclusion, the Cayambes were able to protect their limited resources.

Exercise 1.14: Think of a contemporary group that attempts to practice exclusion in order to protect itself. Does exclusion keeps corruptive influences out? Explain.

_____

_____

_____

_____

_____

For the Cayambe, exclusion seems to have been a useful strategy during the last period, the Inca Period. This exclusionary philosophy might have enabled the Cayambe to resist Inca domination for a long time. Remember, this is one of Sam, Ana, and friends' biggest questions: why was it that the Cayambe were so good at resisting the Inca when the other cultures fell to them so easily?

The Cayambe Inca Period

The Inca Period begins with the Inca conquest and ends with the Spanish invasion. This time period dates from 1505/1525 A.D. to 1534 A.D. Clearly the Inca's glory days in northern Ecuador didn't last long. The beginning dates are variable because it's unclear when this area was finally occupied by the Inca (it took anywhere from 8 years[12] to 17 years[13]). If mounds represent the Late Period, then the feature that best characterizes this era is the hilltop fortresses, sometimes called *pucará*.

Excavations and research conducted by PAP have shown that while many of the fortresses at Pambamarca belonged to the Inca, other *pucará* were probably of Cayambe origin. In general, the fortresses located at higher elevations (over 10,500 feet) tend to be Inca, while the lower ones (below 10,500 feet) seem to be Cayambe (see Figure 1.4). Inca fortresses were built to defend against attacks from the northeast (where the Cayambe lived), and Cayambe pucarás were meant to repel assaults from the southwest (from Quito, under control of the Inca).

Exercise 1.15: We told you earlier that archaeologists have to use artifacts to reconstruct what life was like for ancient people. Below are some facts that we believe to be true about the Cayambe. Describe the archaeological signatures that inform these beliefs.

We believe that the mounds belonged to the Cayambe. What building materials might help us come to this conclusion?

_____

_____

"I knew nothing of the *cangahua* and fortresses or even the type of architecture present in the area. Working at a site like Pukarito [a Cayambe site] also reinforces that you have to keep an open mind and not make conclusions until you have all the information in front of you. *Cangahua* architecture rocks!"

—Siobhan Boyd, 35, of the Gardiner Museum

"The PAP study area is immense with multiple sites, and it can be a little overwhelming, but I thought it would be an amazing opportunity for a dissertation project. I'm interested in the archaeology of warfare and state expansion."

—Matt Schauer, 27, of the University of Illinois at Chicago, Ph.D. student

We believe that the Cayambe had centralized government. Why?

_____

_____

We believe that the Cayambe were preadapted to resist invading forces. Why?

_____

_____

THE INCA

You have to understand the approach the Inca took to imperialism in order to appreciate the role they played at Pambamarca. The Inca Empire was gigantic. It extended through the length of South America, contained 1,000,000 square kilometers of territory, and was home to 10 million people. Despite its impressive size, it was a short-lived phenomenon lasting perhaps just shy of one century.[14] Nonetheless, by about 1400 A.D., a group of 100,000 people had taken over land belonging to 100 times as many—and it took the Inca only a few decades to do so. *Tawantinsuyu*, the name the Inca gave to their kingdom, was the largest political unit known to the Americas.

Such rapid colonization is extremely rare. Scholars have debated how and why the Inca were able to accomplish so much in such a short time. There are two main hypotheses regarding the reason for the unsustainable growth. The first idea is that the bigger the Inca kingdom became, the more resources it needed. The Inca therefore had to conquer more people in order to extract taxes to maintain their high standard of living. The other hypothesis has to do with the way Inca royal succession was decided. It was based on the merit of the legitimate heirs, not birth order. In other words, the worthiest son, not the oldest, would get the crown. Surely gaining more square miles of land in the name of the Inca was a great way to show Dad who should get to be the next Inca king.[15]

The secret to Inca imperial expansion is that they didn't trifle with winning the hearts and minds of the people they conquered: they only wanted their money. The Inca coerced local groups into paying their taxes to the Inca, but the local people generally continued their daily business as they always had. They lived in the same houses, worshiped the same gods, farmed the same fields, and paid tribute to their chief—just like normal. The main difference was that their chief was an Inca. The Inca didn't insist on controlling every aspect of people's lives, which made domination far easier to accomplish in less time. As long as the local people pledged allegiance to the Inca, the Inca were satisfied and the locals were largely left to their own devices.

The Inca were merciful to groups that surrendered, but they were cruel to those who resisted—and their reputation usually preceded them. When the Inca threatened war on a community, the community often relented, knowing they would almost surely be defeated. Of course, not every group would roll over so easily, including our friends the Cayambes. Inca scholar D'Altroy[16] explains:

> The success of the Incas' diplomacy elsewhere similarly hinged on their ability and willingness to crush resistance. In part to deter rebellions and perhaps to make examples, they massacred some especially obdurate foes, such as the Guarco of the coastal Canete and the Cayambes of Ecuador.

Generally, Inca colonization wasn't about revolution; it was about territorial expansion. They just wanted indigenous communities to pay their taxes and not act up. However, the Inca did have one policy that would have caused heavy social disruption: the arrival of *mitmajs*. *Mitmajs* were loyal Inca immigrants sent to populate newly-conquered parts of the empire.[17] This policy helped to minimize local uprisings and was used in the most contentious territories. The Inca believed that if the local population was left alone together to fume about being conquered, eventually there might be a backlash. Alternatively, if faithful *mitmajs* were added to the population, they might serve to diffuse any potential insurgencies.

Exercise 1.16: What type of evidence might archaeologists find to verify the existence of *mitmajs* in a given territory?

_____

_____

_____

Pambamarca Battles and Settlement

The Inca empire's northern boundary stopped just shy of Cayambe lands, and the Inca wanted to add this region to their kingdom too. There were many reasons for wanting Cayambe territory. First, they had prime agricultural lands, so there was a potential financial benefit. Next, Cayambe land was part of a lucrative trade route to the Amazon. Finally, Cayambe land is literally on the equator, and natural landmarks were important to Inca religious beliefs.

Whatever their reasons were, there's no doubt that the Inca put a lot of effort into pacifying the Cayambes. An academic studying the Inca explains that, "The complex of . . . Pambamarca . . . was by far the most extensive array of strongholds concentrated in one region."[18] The strength of the Ecuadorian resistance against the Inca is shown in how many forts the Inca built in northern Ecuador as compared to the rest of their kingdom. You don't waste time and expense building forts in a distant hinterland unless you really need them. And you only need them if you're experiencing significant insurgencies from local communities.

Spanish chronicler Cobo reports on the difficulties the Inca had conquering the Cayambe at Pambamarca.[19]

"The Cayambes, particularly, being men of valor and courage, made it so difficult for the [Inca king] Wayna Qhapaq and his captains that in conquering them a great deal of time and blood were lost. The [king] undertook this conquest in person with a very powerful army, whose commander was Apu Cari, a lord from Chucuito; the [king] entered the land of the Cayambes, carrying the battle forward with fire and blood. Finding that their forces were not sufficient to face the [Inca] on an open battlefield, the Cayambes withdrew and made strongholds in a very large fortress that they had; the [king] ordered his men to lay siege to it and bombard it continuously; but the men inside resisted so bravely that they forced the [king] to raise the siege because he had lost many men in the assaults on the fortress.

Sensing that the opposition was weakening, the Cayambes came out to meet them and they pressed the attack so much that the orejones, who were the backbone of the army, broke and fled, abandoning their king; in the confusion of his men, who were fleeing wildly, the king fell to the ground [out of his litter—the bed they carried him around in], and if the captains Cusi Tupa Yupanqui and Guayna Achache had not helped him and removed him from danger, he would have died at the hands of the enemies. The [king] ordered that, before combating the castle again, his men should make war on the surrounding towns, so that, being deprived of the help coming from nearby, the besieged men would surrender; and, leaving captains to carry out his order, he returned to Tomebamba, where he refused to enter on his litter as he usually did, but entered on foot, in front of his army, with a dart in one hand and a round shield in the other."

This type of insult to the Inca king wouldn't happen again. Now it wasn't only about imperial expansion, it was about revenge. The next battle was decisive for the Inca. Cobo[20] describes the fight.

"The [Inca king] was upset about this loss, but he brushed it off saying men were the food of war and he was only upset about his brother's loss. After mourning, he formed a more powerful army by calling in reinforcements from Cuzco and enlisting other highland Indian groups to help. He sent one general to sweep around one side of the fortress and another general to come around on the other side, destroying anything in their wake (as far as reinforcements, resources went) and then meet with the Inca's men. They did this undetected as was planned. The initial fighting lasted a few days until the [king's] men went in false retreat. When they did this, the Cayambe came out of their fortress and the [king's] ambush troops met them. The Cayambe were overwhelmed by this trick and they laid down their arms and tried to hide near a large lake located nearby. The [Inca king] was intent upon revenge so he ordered his men to kill all the Cayambe they could find. Many men were killed and thrown into the lake until its waters ran red. It is called the 'lake of blood' [Yaguarcocha] even today (Figure 1.7)."

According to historical resources, once the Inca finally overpowered the Cayambe, life as these people had known it would be forever changed. Although the Inca usually left newly-acquired territories alone, because the Cayambe were so difficult to defeat, the Inca wanted revenge. Therefore, unlike other places in the Inca kingdom, the social upheaval for the Cayambes was extreme after Inca domination.

The first major change was that the Cayambe population was seriously depleted due to the Yaguarcocha massacre. Because the Inca had such a difficult time defeating the Cayambe, they had to bring in many *mitmaj* to ensure the Cayambe region remained loyal to the Inca Empire. This has had an effect on the linguistics of this region today. In small communities in the Andes (like in the village of Cangahua), many of the locals speak Spanish and Quichua (the language of the Incas). The original Cayambe language is lost to history forever.

Exercise 1.17: How do you think life for the Cayambes would be different today if this Inca king hadn't been born? Do you think someone else would have taken his place? Why or why not?

_____
_____
_____
_____

Exercise 1.18: Think of a historical or contemporary person who has single-handedly impacted the world. How would the world be different if this person hadn't existed?

_____
_____
_____
_____

FIGURE 1.7 Lake of Blood. Now, a tourist destination. Courtesy of PAP/Samuel Connell.

Obviously this historical account is pretty dramatic. Wars, kings, blood, battle, massacre, and social upheaval are attention-grabbing. The intrigue of archaeology is often discovering the pathos of the ancient world. But remember, we're archaeologists, and we need to keep our evidence in mind. Often, we'll look for verification of historical accounts or local lore through material culture. We told you earlier in this chapter that the Cayambes built forts of their own and possibly used them to repel the Inca. How can we tell archaeologically if the forts at Pambamarca are Incan or Cayambe?

Understanding cultural patterns (like what we practiced in the Group Exercise) is critical for figuring out who constructed which fortresses. If the Inca typically use a different type of building material than the Cayambe do, examining building materials might help us decide who built the fort. Perhaps the two groups used distinct pottery or had dissimilar weapons. Or maybe they preferred to build in landscapes that weren't alike. If we can learn about these types of patterns, then it helps us make reasonable assumptions about the culture group using a given structure.

For instance, you were told about Quitoloma when this chapter began. Quitoloma is the quintessential Inca fortress (Figure 1.8) because it contains all of the typical Inca patterns (studied in other parts of the kingdom). First of all, it is made out of locally quarried stone. The Inca are well-known masons and prided themselves on their superior

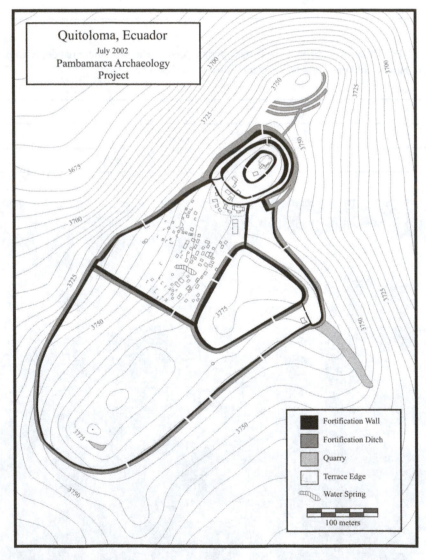

FIGURE 1.8 Quitoloma map depicting concentric circles and gates. Also note the buildings inside the fortress. Courtesy of PAP/Samuel Connell.

stonework.[21] Next, Quitoloma has a large earthen platform at its summit that is believed to be an *ushnu*. *Ushnus* were Inca places of worship. Finally, Inca pottery and *bolas* were found. *Bolas* are sling stones, which were a common weapon used in Incan wars. All of these patterns indicate that Quitoloma is, indeed, Incan.

Comparatively, archaeological research at other Pambamarca fortresses indicates that some of them are Cayambe in origin. One such example is called Pukarito. For instance, its walls were built over the course of three construction periods. The earliest evidence of construction was started well before the Inca invasion. A radiocarbon sample taken from an excavation confirms that Pukarito has been around since the year 1400. The Inca didn't show up until at least 1505. Furthermore, the fortress was made out of *cangahua*, the preferred building block of the Cayambe.

Exercise 1.19: What is the typical style of house or building where you are from? What materials are used to build them? Why? How do the architectural styles from various parts of the country differ?

_____

_____

_____

_____

_____

As you can see from the previous exercise, paying special attention to typical building materials, architecture features, and geography really makes a difference when determining which cultural group might have built structures or used artifacts. Distinguishing who owned the fortresses can ONLY be known after a full accounting of the history of the area. Some students can't wait to get the trowel and start digging. (Can't blame them, that's the fun part!) But if you have no understanding of the groups or their materials, you'll not understand the significance of what you're discovering in the ground. Until you understand what to look for, all of that digging might just make you dirty, not more knowledgeable.

THE SPANISH ARRIVE

The story of the Spanish at Pambamarca begins with the fall of the Inca in 1534. Christopher Columbus landed in the "new world" in 1492, just before the Inca finished their imperial expansion into northern Ecuador. It wouldn't be long before all indigenous people of the Americas were under European rule. In 1532, Spanish conquistador Francisco Pizarro and his small band of 62 horsemen and 106 foot-soldiers marched into the Peruvian interior and had almost immediate success overthrowing the Inca kingdom.[22] As history shows, coincidental hardships for the Inca like being in a civil war when white men appeared on their shores enabled a few hundred Spanish soldiers to take over an empire. The technological advantage of horses and steel worked in the Spaniards favor as well.

Pizarro's arrival in South America couldn't have been better timing, from the Spanish point of view. The Inca population, resolve, stores, and society had been devastated by civil war. Pizarro also had a weapon of mass destruction on his side, even though he wasn't aware he had it. That weapon was disease. Old World diseases like small pox were rapidly spreading through Central America into South America. Some estimates suggest that as much as 95 percent of the local population was wiped out by these illnesses.[23] Pizarro's last bit of good luck is that he and his men coincidentally ran into the victorious Inca king returning home from the civil war and promptly killed him. As an Inca scholar[24] rightly observes, "The Conquest began with a checkmate."

Hacienda Period

After the Spanish conquest, the Crown distributed wealth found in the New World to its conquistadors. This is what created the hacienda system, which is similar to the antebellum plantation system in the American South. Being a conquistador was the only way a Spaniard born of humble beginnings could attain wealth in a rigid class society where the luck of one's birth would often determine a person's earning potential for life. Offering conquistadors money ensured that the Spanish Crown would have trustworthy residents living in its new territory.

The Spanish were like the Inca in this regard. Like the Inca *mitmajs*, the Spanish knew the only way to avoid uprisings in a freshly conquered land was to install their own loyal citizens. When the Inca treasuries were depleted, the Spanish Crown had to find a way to repay its conquistadors and convince Spaniards to remain in a foreign land.[25] The answer to this problem was *encomiendas*. These were the predecessor of the hacienda system.

Allotments of indigenous people ('indigenous' to the Spanish were Inca or Cayambe . . . they didn't differentiate) were given to a conquistador to assist with farming, domestic duties, etc., for a few months out of the year in exchange for religious instruction. This system was called *encomienda* and was meant to reward the conquistadors with profitable land and Indian laborers, as well as attract other Spaniards to settle on the new continent. Spanish citizens populating the new world ensured stability for the Crown, but the indigenous were unwillingly forced into a feudal system (i.e., they never asked for religious instruction or to be slave laborers).

A Spanish chronicler paints[26] a grim picture of their life:

"This manner of payment is far worse than if they paid them nothing but kept them branded slaves in their houses: for a master gives his slaves food and clothing and cures him when he is ill. But they make an Indian work like a slave and give him no food or clothing or medical attention: for the maize is insufficient to feed him and the salary not enough to clothe him."

The main difference between a hacienda and an *encomienda* is that the *encomienda* was a system where Indians helped out the Spanish land owners with the laborious duties of owning large swaths of land. The *encomiendo* was not earning a profit, per se. It was just a life of luxury. A hacienda by contrast was its own little economy. The haciendas produced something (like wool or textiles made at the Hacienda Guachalá) that could be sold for a profit. It was on the backs of the indigenous that the Spanish hacienda owner not only survived off of his land, but profited from it. After all, labor costs are zero when you employ slaves or indentured servants. Frequently the Spanish *encomiendo* would hire a local person (sometimes an ethnically mixed mestizo Spanish/indigenous person) to be the foreman at the hacienda. This person would organize labor and often very cruelly mete out the owner's wishes, often inflicting hardship on the workers for the Spanish overlord.[27]

Hacienda Guachalá

The history of the Spanish at Pambamarca is best expressed by the Hacienda Guachalá, which is in PAP's territory. Francisco Pizarro himself laid the foundation for the existence of the Hacienda Guachalá by naming fellow conquistador Pedro Martin the *encomienda* of Cayambe in 1535 (the Hacienda is currently in the county of Cayambe). The Hacienda is named after an Indian *mitmaj* named Guachalá who was brought to the Cayambe region after Inca domination. It was a large property, and soon the first house and church were built.

Guachalá eventually became a significant place in the Pambamarca landscape. In 1700, the hacienda got work licenses to start manufacturing wool on the premises. This is the same year the first room of the *obraje* (work house) was built. Hacienda Guachalá came to export textiles made on the property to Lima, Bogota, and Spain.[28] Once Guachalá started producing wool, etc, for profit, it became part of the hacienda system.

The first inventory of Guachalá was made between 1767 and 1771, which described what the place looked like. Most of the roofs were made of straw, the doorways to the house were covered in cowhides. The central patio looks like it does now, though the cobbling was not present (Figure 1.9). Select buildings had tile roofs and wooden doors. The entire estate was enclosed by a high wall made of *cangahua* blocks about a meter thick. Also on the property were Kashmir wool, dyes, caldrons, and water systems, which indicated an increase in production level at the property. In 1779 the church at Guachalá was the main one for the region.[29]

Sometime between 1819 and 1824, another inventory was taken. Sadly for the indigenous (including those of Inca decent) people who were forced to work at the hacienda, a dungeon had been built. According to early documents, slaves were also present on the property. In 1892 the first member of the Bonifaz family owned the property (the Bonfiaz family owns this hacienda today). The workers were granted small plots of land to live on and plant their food. This left them in perpetual debt to the hacienda. New land reforms from 1963 to 1973 forced Guachalá to sell off most of its 12,000 hectares. A second church (Figure 1.10) was built on the property in 1938, and the torture room was destroyed in a fire in 1963. Hacienda Guachalá has been a hotel since 1993.[30]

Exercise 1.20: Historical records describe what the Hacienda Guachalá looked like in the past. It is still standing today. What might you expect to find in the archaeological record that could confirm the historical documentation?

_____

_____

_____

_____

_____

FIGURE 1.9   Hacienda Guachalá courtyard with cobbles. Note roof made of tile.
Courtesy of PAP/Samuel Connell.

Because it was founded in 1535, this hacienda has been affecting the settlement patterns in the Pambamarca region since less than ten years after the Cayambe were subjugated by the Inca. The history of the Spanish in South America and the emergence of the hacienda system are important to this project. It speaks to questions about land use and dominance by a second imperial force in the Cayambe territory.

FIGURE 1.10   Hacienda Guachalá chapel.
Courtesy of PAP/Samuel Connell.

"The reason Ana and I wanted to dig at Pamba-marca is because the sheer size of the fortress (pucará) was stunning—the labor invested and the mystery behind their construction was amazing. But as time went on, we became more interested in the story of resistance.

The other thing about archaeology is that choosing where you want to study, excavate, survey, or analyze depends on lots of factors, and sometimes they're personal. Here are the rest of my reasons. Think of what might motivate you to find out more about another place.

After architecture, another decisive factor was that Ecuador was wide open, and the country seemed to thirst for foreigners to come. It wasn't already overrun with archaeologists. Here we had a chance to make a real contribution to the understanding of the culture history of the country.

From another perspective, it's been fun to work with my wife in her home country. Bringing our children is incredibly rewarding, and getting to know her family has been wonderful.

Also, the growth of field school students into graduate students has been awesome.

But, best of all has been the daily life with the people of Ecuador. It's something to behold, and nothing I want to give up. The smiles and feelings of connections (and sometimes disconnections!) to the people of Ecuador has been the strength behind the project from the start in 2000."

OPTIONAL RESEARCH OPPORTUNITY

INSTRUCTOR'S NOTE: This is an option to expand the discussion of topics raised in this chapter.

A topic of this chapter has been about what prompts an archaeologist to study in a specific region. Therefore, this take-home research opportunity lets you explore how current events, contemporary interests, or even pop culture influence what an archaeologist finds interesting.

The National Geographic magazine contains articles about archaeology and popular culture. Hypothetically, the magazine could cover any contemporary archaeology project (or anthropology issue) taking place in the world, but it doesn't. The National Geographic covers stories about nature, history, or people based on current events—topics that are currently in the public consciousness. For instance, they ran a cover story on Charles Darwin in February of 2009. Coincidence? Think again! That was the 200th anniversary of his birth.

Therefore, for this research opportunity, you need to find at least TWO National Geographic articles from TWO separate decades. In a concise one page essay, explain what drew your attention to the two articles. Then explain why you think the magazine chose to cover the particular stories based on current events of the time. Compare and contrast the articles by brainstorming some of the ideas below.

1) How do archaeologists reinterpret their hypotheses about culture and social changes based on the contemporary ethos of the time in which the archaeologists lives?
2) Check out other articles and advertisements in the issue. How does this additional material help contextualize your article? Are they a "sign of the times"?
3) What about the stylized artwork done by Geographic artists? Does it change through time? How about the photographer's subjects? Does the art reflect the values in the articles too?
4) Some say that archaeology will only survive as a field as long as it proves itself pertinent to society today. How have archaeologists endeavored to remain pertinent in the eyes of the public?

Chapter 2
Fieldwork

In the last chapter, we told you about the culture history of the Cayambe, Inca, and Spanish. We talked about what food they ate, what kinds of houses they built, and the way they organized themselves. Some of this information came from historical accounts, but a lot of it was discovered during archaeological fieldwork. This chapter describes how archaeologists recover data and interpret it in the lab.

Archaeology can be grueling, tiring, and dirty work. You might have to hike up steep mountains in the high altitude air, work in grimy clothes, and deal with the occasional tarantula. Fieldwork is also exhilarating. It is often students' favorite part of field school because it's the ultimate discovery process. You get to find something in the ground that has been lost for centuries. The things people dropped, left, buried, or built long ago will help them speak to you—tell you about their lives and humanity's past.

In this chapter, we'll deal with three main elements of archaeological fieldwork: survey, excavation, and the lab. Entire volumes have been written about the methodology of excavation alone, so this chapter won't cover every detail. However, it should be enough to give you a taste of the process. We'll show you some real life examples and let you do analysis on your own. Archaeologists are the detectives of history, and in this chapter, you get to be the sleuth.

STEP 1: SURVEY

Survey is often considered the first step in archaeological fieldwork after library research is complete. Archaeologists use surveys to assess what types of sites, features, structures, or loci (places used by humans) they have at their project area. This is the discovery process, when archaeologists determine what places in a given area might have been used by humans.

When archaeologists conduct surveys, they try to identify places in the landscape that humans used in the past. Quitoloma is a great example of what might be found in a survey— a building not previously surveyed or mapped (or done so in recent times). Quitoloma is a massive architectural site, but other sites could be buried or more ephemeral in nature— such as a battleground. Surveys can identify the places that humans have utilized, modified, built on, or altered. Maps must be made of these places so archaeologists can visually see how they relate to one another.

It is crucial to determine the boundaries of a site and how various sites interconnect. For example, the boundaries of Quitoloma might extend far beyond the confines of the fortress walls. Were other armies camping outside of the fortress? Were suppliers or salespeople there? Were there auxiliary forts or forward operating bases elsewhere on the landscape that supplied or supplemented Quitoloma's activities? If the archaeologist only identifies the largest (or most intact) buildings, he/she might miss the other associated features that would make interpretation of such places more clear.

Survey commonly includes the following methods: field walking, sampling, shovel testing, aerial photography, geophysical survey, and mapping. The type of surveys conducted will be based on the research question being asked and archaeologist's time and budget constraints. If someone is interested in Inca fort construction, that person will spend extra time documenting Quitoloma's architecture in detail. By contrast, other archaeologists might be interested in locating individual artifacts, while others want to know about vegetation or soils. Other people might want to find places mentioned in ethnographic or historical documents. These individual research goals will help determine how and where an archaeologist approaches fieldwork.

TABLE 2.1   Various Types of Archaeological Surveys.

SURVEY TYPE	PURPOSE	ADVANTAGE	PROCESS
Field walking	Record sites that you literally walk past; find artifacts on the surface of the ground.	Simple procedure; good for beginners.	Have people spaced evenly along the landscape; record each cultural feature along the transect they're walking.
Sampling	Record findings (like artifacts) discovered in a small sample of a large area; extrapolate the results.	It's a small enough process to reasonably be completed in a field season.	Some areas are too large for every square inch to be recorded in detail. Instead, a small sample is field walked or otherwise surveyed. The results are then assumed to be statistically representative of whole area.
Aerial photographs	Save your aching legs and identify structures from photographs taken from the air where you can see everything at once.	See features too big to notice when field walking (imagine a 10-mile long wall, for instance).	Examine crop marks (places were crops won't grow) or soil marks, which could indicate a site not visible from the ground. Can be used when a large-scale view is needed to "see" the site.
Shovel testing	Similar to field walking, but you dig a shallow hole every few paces to understand what might be lurking just below the surface.	It applies the principle of sampling to the subsurface—it lets you see a little sample of what is underground.	Everyone walks along their transect line. However, every 100 paces (for example), each person digs a shallow hole and records any cultural material.
Geophysics surveys	Find underground sites.	Not every archaeological site contains a standing structure. This is a way to locate such sites.	Usually involves scanning the ground with a machine. The machine gives geophysical data that you can use to locate subterranean sites.
Mapping (GIS)	Record archaeological info for future analysis.	With GIS, a user can do searches with a map; for instance, view only Inca artifact finds, etc.	Input survey/excavation data into a computer program so that various maps can be created with the touch of a button.

"Favorite part of fieldwork: survey because you get to see a lot. Least favorite: trying to lay a grid at the top of a ridiculously windy mountain at freezing temperatures in the rain. The why should be self explanatory."

—Erin Rodriguez, 21, University of Pittsburgh, Junior

Table 2.1 shows some of the more common types of surveys. It also explains how they're used and why. As you can see, there are many options to choose from. Some archaeologists use all of these methods at different times during a project. How do you choose which survey is best? It depends on the goal of the research and the local landscape. For instance, if an archaeologist wants to know how fortresses relate to each other over many square kilometers, that person will probably not do shovel tests. Fortresses are big and are visible on the surface of the ground. Digging a tiny hole every 50 paces isn't likely to help you find a gigantic building. Choosing the proper survey method depends on what will best get the job done.

However, the goals of a project might evolve. Once all of the fortresses have been located, perhaps the researcher would like to know if people lived on top of them. If there are no buildings, the soldiers still could have lived there in tents. Of course, you aren't likely to find 500 year old tents. In this scenario, a shovel test survey would be appropriate. A small hole could reveal domestic artifacts that lie just below the surface. Domestic artifacts could be food remains or pottery. These would indicate soldiers had indeed lived on top of the fortresses, even if they didn't reside in a house. In this scenario, the archaeologist might first field walk and then shovel test in specific areas.

For the following exercises, think about the methods listed in the table. Given the following scenarios, explain which method(s) you would use and why. Use the table to help you decide.

Exercise 2.1: You are a field director with lots of new students. You've just been given permission to add five square miles of land to your potential project area. You've never been to this land before, but according to the locals, ruins are there. What method would you use for survey and why?

_____

_____

_____

Exercise 2.2: You are searching for pre-Inca sites. Unfortunately, the Cayambe mound builders didn't actually build mounds in the PAP project area. Despite the lack of pre-Inca structures on the ground surface, you need data. What kind of survey might you use to find where they lived and why?

_____

_____

_____

"Although the landscape and environment was different some 500 years ago, it is sometimes hard to imagine how these people could have lived in this environment, which was a war zone that lasted 9–17 years."

—Amber Kling, 29, Ph.D. student at State University of New York at Buffalo

OPTIONAL RESEARCH OPPORTUNITY

Exercises 2.1–2.3 are easier and more fun if you consult Google Earth. Get on Google Earth at home and type in some of the areas we've discussed in this book so far. Try Hacienda Guachalá, Oroloma, and Quitoloma. What do you see? This is a great example of the use of GIS. You can see how these sites relate to each other and what they look like from the air. What can you see from the air? What can't you? How does looking at Google Earth inform your choices in the previous exercises?

Exercise 2.3: You have found eight separate fortresses over a 15 square mile area. You have noticed that there seem to be trails or ruts that lead from fortress to fortress, but you can't tell if these paths connect every single fortress to the other. The paths are hard to see, the distance is great, and field season is almost over. What kind of survey might you use and why?

_____

_____

_____

Archaeologists might be detectives of prehistory, but they're also business managers. Like a manager, the archaeologist has budget restrictions and must consider overtime for his people. Though it would be wonderful to do very detailed surveys of every archaeological project, that would be difficult to accomplish in real life. At least, not in one field season!

Geophysics Surveys

Most of the surveys we've discussed assume that you can see artifacts, structures, or cultural remains on the ground. Although many people get interested in archaeology because of the amazing prehistoric temples, pyramids, palaces, and fortresses that were built in the past, the truth is, many structures do not stand the test of time. Lots of things can ruin a building: new construction, water or wind erosion, purposeful bulldozing, or general neglect. Sometimes only the foundations of former structures are left—or maybe humans were using places that didn't necessarily require a building (like a park or tent). Most archaeological sites have no sign of existence on the surface of the ground.

Though shovel testing provides a peek underground, remote sensing can show the dimensions of buried structures and how they relate to each other over a large area. Instead of merely glancing at a small percentage of the subsurface (as one does with shovel testing), remote sensing "fills in the blanks" and provides a map of it all. For example, shovel tests might prove there is a wall underground, but it could be difficult to determine the size of it—especially if it's large. Remote sensing can show the boundaries of that wall and even how other buried structures intersect with it. Remote sensing means "from far away" (like aerial photography, for instance). "Remote" can also mean using geophysics tools to test disturbances in the soil that might indicate where underground sites are located.

Though geophysical survey methods are becoming more common, they're still not in mainstream use in America. The equipment is expensive and requires specialized training to operate optimally. Additionally, it's best used in conjunction with dedicated software, some of which requires an expensive subscription to use. PAP is lucky to have access to a variety of geophysics tools. However, we'll limit our discussion of geophysics at Pambamarca to GPR—ground penetrating radar. It's the method used most prominently at this project area.

GPR at Pambamarca

A great example of the use of GPR at Pambamarca comes from the site named Oroloma, which means "gold hill" (so named because of a local legend about buried treasure discovered there). Today Oroloma is a bean field. When the soil is tilled, artifacts are churned up on the surface of the ground: pottery pieces, stone tools, and even bones. All of these artifacts hint at what lies below. Some tests on the artifacts indicated that this is a very old Cayambe site. It dates back to the year 700, far before any of the Inca or Spanish battles.

Sam and the directors knew they wanted to dig at Oroloma, but the problem is that Oroloma is a large space—many square acres. And, there are absolutely no structures, foundations, or building blocks whatsoever on the surface of this field to point to where, exactly, the biggest concentration of artifacts or sites would be found (Figure 2.1). The farm machines till the soil, thereby spreading cultural material all over the site. How will the archaeologists know the best place to dig?

In a short summer field season, efficient use of time is critical. Trial and error excavations are wasteful, costly, and tenuous. For instance, when the farmer is ready to sow his seeds, archaeologists have to be gone regardless of the stage of discovery. An efficient way to determine where to dig would be to first conduct a GPR survey. This survey could show the archaeologists the location of any buried structures or buildings.

How It Works

GPR works by emitting an electromagnetic wave into the earth via an antenna that is dragged along the surface of the ground. These waves are produced by the acceleration of an electrical charge that is sent into the ground by the antenna. The transmitted wave will encounter substrata (i.e., layers of dirt) along its journey. If the wave hits different types of substrata on its journey through the ground, it will reflect off of those changes in the soil and bounce some waves back to the antenna. A computer is attached to the antenna and records the various reflections of the electromagnetic wave. Figure 2.2 shows the components of the machine.

FIGURE 2.1   General view of Oroloma showing slightly undulating empty field. Courtesy of PAP/Samuel Connell.

Computer that records
radar signals

Antenna that transmits radar signal

Survey transect line

Meter marks along transect

FIGURE 2.2   Picture showing components of GPR and how survey is conducted along carefully marked transects, with each meter of the transect marked so it is clear how the data relates to a real place on the ground.

An example is the following. The radar wave will move through homogenous soil at a certain rate (meters per nanosecond—that is meters per billionth of a second) until it hits something different, such as a buried wall. When the radar wave encounters the wall, it will either speed up or slow down. The computer records how the wave changes when it reflects against something different (like a wall). That underground change is called an anomaly because it is something out of place. An anomaly may indicate the existence of architecture or a group of artifacts, for instance. This tells the archaeologist where cultural material is potentially located, and he/she will then know where to dig. (Note that not all anomalies are cultural; this will be discussed later.)

Look at Figure 2.3. You can see a picture of two people surveying. The gray block beneath them is the data they collected as they surveyed one single transect (line) in their survey area. This gray block of data is called a *radargram*, and again, it represents a single walked transect and all the ground beneath their feet along that transect. In this case, you can see from the scale that the radar penetrated 200 cm below the surface before losing steam or petering out. (The type of soil and conditions surveyed in determine the distance the wave can travel underground.) The gray block is what they will see on their computer screen. Anomalies are pictured as places that look different from the rest of the data—maybe a darker or lighter area, a place that appears squiggly, or in a V or X pattern, etc.

What is shown immediately under the antenna in Figure 2.3 is an anomaly. It looks different than the rest of the soil because it is something out of place. When the radar wave encounters this anomaly, it sends a signal back to the antenna that something out-of-place is located at approximately 100 cm below the surface of the ground. The wave then continues to make its way through the ground, beyond that anomaly, until it can travel no farther. It then goes back to the antenna and the data is recorded by the computer. GPR can record multiple anomalies above and below each other with the depth of those anomalies given, as shown in the Figure 2.3 example.

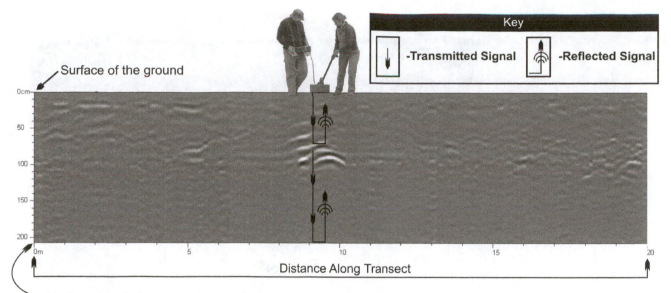

Surface of the ground

Maximum depth that the radar signal
reaches under the ground

FIGURE 2.3 Picture showing the type of data collected for each transect in a survey. This vertical slice is called a radargram. The picture also shows what an anomaly looks like in the data, and how anomalies are recorded by the antenna and computer of GPR.

GPR surveys are conducted by dragging the antenna along a series of surveyed lines (transects) within a survey grid. Look at Figure 2.4. It shows an exactly measured grid with survey lines spaced one meter apart. The archaeologists will drag the antenna along those one-meter spaced transects. They will make a mark at each meter along the transect. It is important to be precise about data collection because this is when the x, y, and z coordinates of an anomaly are being recorded.

The archaeologist has to know where anomalies in the data correspond to real places on the ground, so he/she knows where to dig—hence why these carefully marked surveys are crucial. Please note that Figure 2.2 shows the meter marks along a transect (at the bottom). When those marks are made while surveying, they show up in the data. For example, see the meter marks at the bottom of Figure 2.3. This is how the distance along the transect is known.

The radar doesn't penetrate the ground as a pinpoint . . . that is, the radar wave is wide, and it will collect data between a half a meter and a meter (depending on the soil) on either side of the transect line. Therefore, by surveying only every half or full meter, the archaeologist has in fact collected data for the entire grid. He/she knows what the soil under the ground looks like

FIGURE 2.4 Survey grid showing individual transects. The GPR was dragged along each of those lines, providing data about the soil underneath the ground surface for this whole survey grid.

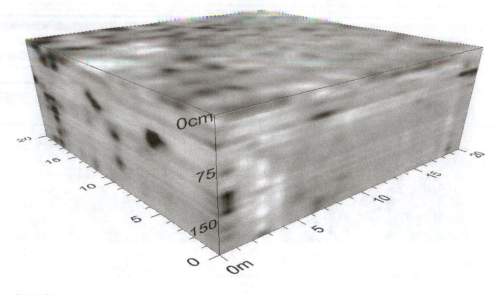

FIGURE 2.5   This cube represents all data from the surface of the ground to two meters below the surface for the survey area pictured in Figure 2.4.

on the x, y, and z axes. Doing a GPR survey gives the archaeologists a cube of data, like is shown in Figure 2.5. Software is used to turn individual transects into this cube.

Another way to envision how the transect lines turn into this cube of data is to look at Figure 2.6. On the right, you can see how each radargram (from the survey transects) is stacked together, thus creating the cube. The software extrapolates what the data is like between transects.

The great thing about GPR is that the software allows the user to view the data in a variety of ways, such as individual survey transects (radargrams, see Figure 2.3), as a cube (as seen in Figures 2.5 and 2.6), or in planview slices. After the software has turned the transects into a cube of data, it can then "slice" it horizontally at various depths, thus showing what the survey would look like at 120 cm below surface, for instance, or at 20 cm below the surface. It's like having x-ray vision in that the archaeologist can pick any depth below surface and

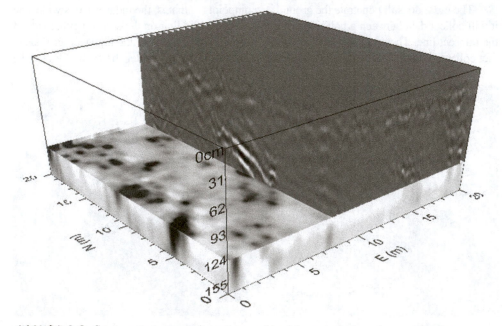

FIGURE 2.6   See how the transects (or radargrams) have been stacked up, thus creating the cube of data. The area on the left that has been filled in shows what it will look like when the transects have all been added in.

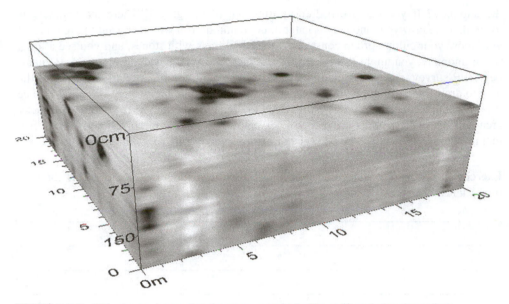

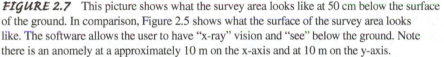

FIGURE 2.7 This picture shows what the survey area looks like at 50 cm below the surface of the ground. In comparison, Figure 2.5 shows what the surface of the survey area looks like. The software allows the user to have "x-ray" vision and "see" below the ground. Note there is an anomoly at a approximately 10 m on the x-axis and at 10 m on the y-axis.

view what the anomalies look like at that depth. Another way to picture what a horizontal slice looks like is to think of an empty swimming pool as it's filled with water. The software allows the user to see the data (or swimming pool) partially filled.

If you look at Figure 2.7, you can see what a horizontal "slice" looks like. You can see that the cube has only been "filled up" to 50 cm below surface. That is a picture of the anomalies at that depth. Note that anomalies in planview look like different colored blobs. The size of the blobs (or anomalies) can be measured on the x and y scale. In Figure 2.7, it's obvious that there is a strong anomaly approximately in the center of the survey area (at x=10 m and y=10 m). An archaeologist would then have the coordinates to know where to start the dig on the surface of the ground. He/she would also know that the anomaly should be encountered after excavating for about 50 cm.

Interpretation of GPR

The beauty of GPR is that the user knows where an anomaly is located and its approximate shape and size. The question is: what is the anomaly? And for archaeologists: is the anomaly cultural? The truth is, no one really knows the answer to those questions until one digs. However, experience, knowledge of the geology of the area, and information about the groups inhabiting any given site can guide interpretation of the GPR data. Paying attention to this type of information can reduce the fruitless digging that comes from excavating at the site of any given anomaly.

We will walk you through how to do interpretation for Figure 2.7, and then later in this chapter you will be asked to apply these guidelines to your own interpretation of GPR data. First, note the size of the survey area in Figure 2.7. It is 20 m wide by 20 m long. This is shown on the scales on the x and y axes. You want to pay attention to the scale because then you know the approximate size of any given anomaly. In the case of the anomaly in the middle of Figure 2.7, it looks to be about 3 m wide by 3–4 m long. Next, take note of how far below the ground surface the anomaly is located. In the case of Figure 2.7, we are looking 50 cm below surface. If you are looking at multiple slices in a row, see how "thick" an anomaly is: does it show up in more than one consecutive slice?

Answering these types of questions is the first step. The next step is to consider where you surveyed, your knowledge of local geology, and what you know about the types of buildings or land use patterns of the area. For example, what might have caused the anomaly? If you can see bedrock jutting out of the ground where the anomaly is located, you know your anomaly is natural, not cultural. You wouldn't want to dig there. If you know your culture group constructed their buildings with stones, and you see a long, linear anomaly, you might assume that it is a wall. You would want to dig there. Remember, GPR survey is not conducted in a vacuum. Use common sense and knowledge of the area to draw conclusions. If you see an anomaly with a right angle, there is a good chance it is man-made since nature doesn't produce too many right angles. If you are in a rocky area and see many boulder-sized anomalies below the surface in no visible pattern, then that might merely indicate the presence of naturally-occurring rocks in the subsurface.

Exercise 2.4: Why is it unlikely that GPR will detect *cangahua* (not the town, but the mud brick) at Pambamarca?

_____

_____

_____

_____

Exercise 2.5: Would a stone paved house floor be easier to see if it was under a layer of mud or a layer or rocks? Why?

_____

_____

_____

Interpretation can also be done with the radargrams. Radargrams have an advantage over the horizontal slices in that the computer highlights anomalies in the slices—and remember, those anomalies can be "false positives." Anomalies are just anything out of the ordinary. With the radargram, you are looking at raw data. If you see an anomaly in these profiles, you can see its shape more clearly. Does it curve with the slope of the land? Does it change shape in a way that differs from the rest of the soil? Also, toggling back and forth between horizontal slices and radargrams can help "confirm" an anomaly in the horizontal slice. Look at Figure 2.8. You

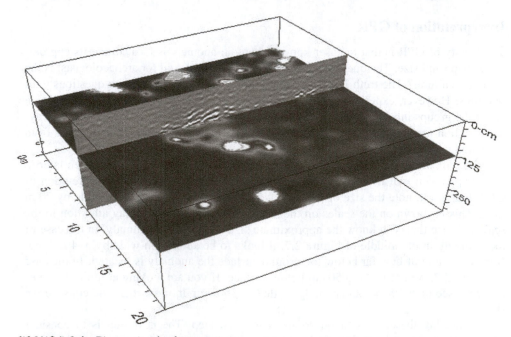

FIGURE 2.8   Picture showing how a radargram relates to a horizontal slice.

can see how the radargram and the horizontal slice relate. If one sees an anomaly in the slice, and it's visible in the radargram, the archaeologist might have found a good place to dig.

Exercise 2.6: Look at these profiles (Figures 2.9–2.11) and see if you can find anything out of the ordinary. Circle what you find. These are anomalies and they are potential cultural sites. Of all the ones you noticed, where would you choose to dig? Circle that place in a different color.

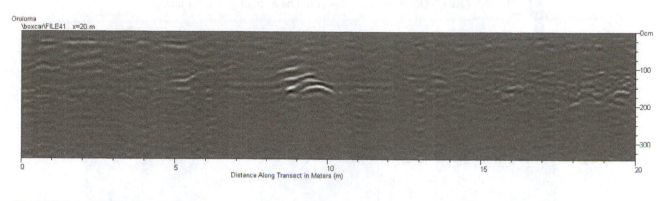

FIGURE 2.9   Showing one radargram from Oroloma, the Cayambe site.

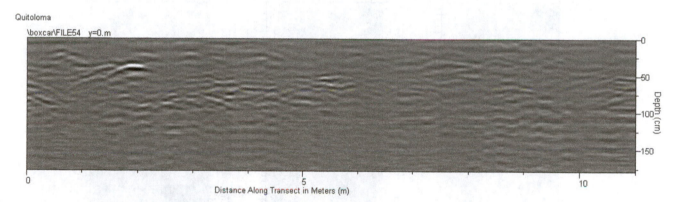

FIGURE 2.10   Showing one radargram from Quitoloma, the Inca site.

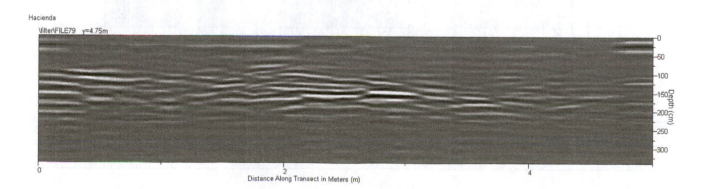

FIGURE 2.11   Showing one radargram from Hacienda Guachalá, the Spanish colonial site.

Exercise 2.7: In this exercise, you will practice interpreting GPR data on your own. First check out the following plan view maps (Figures 2.12–2.14). Circle the anomalies you find. Pay special attention to anomalies that last through multiple slices. Consider their dimensions. Next you'll find a place to write some notes below the figures. Based on what you read in Chapter 1 about the different culture groups, what do you think the anomalies might be? You have some knowledge about what types of sites, building materials, or artifacts each culture group would have used. Use this information to make a reasonable guess about what you're seeing in the GPR data. Write your answers in the lines below Figures 2.12–2.14.

INSTRUCTOR'S NOTE: This could be a small group project.

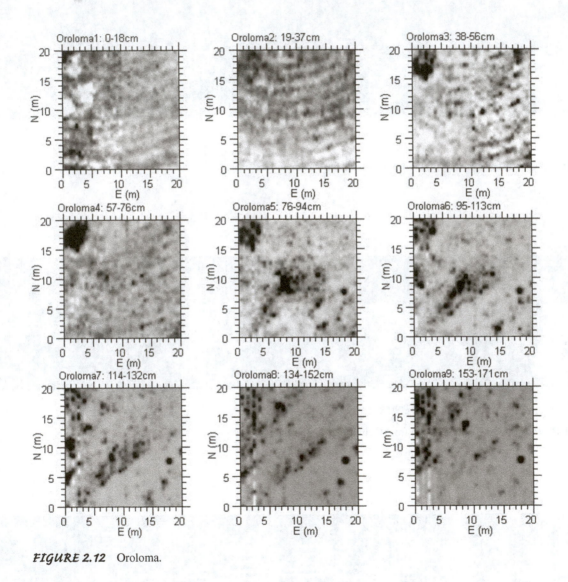

FIGURE 2.12 Oroloma.

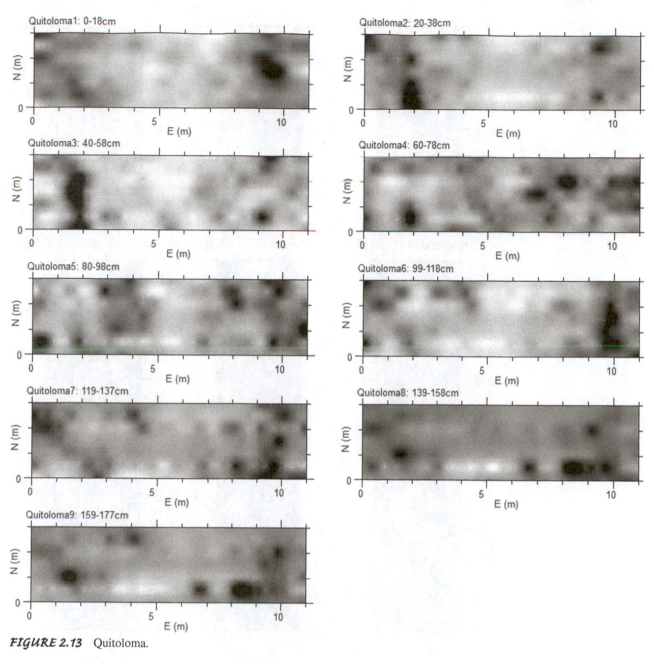

FIGURE 2.13 Quitoloma.

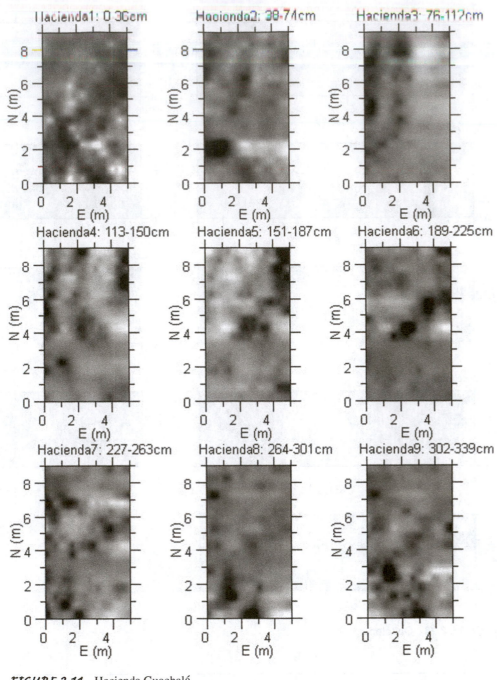

FIGURE 2.14   Hacienda Guachalá.

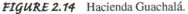

Now that you've finished the previous exercises, you might still feel like there are tons of anomalies. Well, there's another trick up the archaeologists sleeve, and that is relating the research question to the GPR data. For instance, we didn't do these surveys just anywhere at Oroloma, Quitoloma, and the Hacienda. Each survey was conducted to answer a specific research question. The next exercise will help you narrow down the results.

Exercise 2.8: Consider this extra information about the research design at each site. Use it to reevaluate the GPR slice maps in Figures 2.12–2.14. After reading the information below, where would you now conduct your excavation? Use the following spaces to write the coordinates of your dig and the explanation of why you'd excavate there.

File: Oroloma, Quitoloma

Oroloma: A huge bell-shaped pit full of Cayambe garbage was found at Oroloma in the past. It was so chock-full of artifacts that we wanted to find another one. Though we knew it'd be difficult to see *cangahua* architecture with GPR, we knew that a big pit of pottery, tools, and bones would definitely be an anomaly compared to the other dirt.
Oroloma Coordinates:_____
Why there?_____
_____
_____

FIGURE 2.15   The square indicates the location of the *kallanka*. The horizontal line above it is where the mysterious wall is located. Courtesy of PAP/Samuel Connell.

Quitoloma: A mysterious wall was located to the north of the *kallanka*, which was the officers' quarters. We wanted to know if the mysterious wall was once attached to the *kallanka* (Figure 2.15). This would help us understand if the *kallanka* was expanded or added-on over time. This would indicate semi-permanent housing of soldiers, which speaks to the duration of the struggle between Inca and Cayambe.
Quitoloma Excavation Coordinates:_____
Why there?_____
_____
_____

Hacienda Guachalá: We were excavating in what was thought to be the *obraje* of the Hacienda. This is where the workers were imprisoned to make dye or wool. From the looks of the wall in this yard (Figure 2.16), it appeared to once be partitioned. If we could find remnants of a wall, we would know this area was once enclosed and separated. It would help us determine if this was, in fact, the *obraje*.
Hacienda Excavation Coordinates:_____
Why there?_____
_____
_____

FIGURE 2.16   The circled area shows where a wall might have once been. It looks like it was torn off. Courtesy of PAP/Samuel Connell.

"The directors emphasized the importance of taking good field notes, maps, and recording all data. The meticulous nature of archaeology can be frustrating, but it is incredibly important as this information can never be excavated again. I was used to a more casual approach to excavation, which I learned could corrupt the data you find."

—Matt Schauer, 27, Ph.D. student at University of Illinois at Chicago

STEP 2: EXCAVATION

If there's one word synonymous with *archaeologist*, that word has to be *digging*. Isn't that what everyone imagines? Someone in a khaki outfit and hat, bent over in the hot sun, meticulously exposing some clue about ancient man? In fact, excavation is an important part of data collection. Often, it's also an archaeologist's favorite part of fieldwork. It's physical, in the great outdoors, you find the majority of artifacts, and it's fun to get your hands dirty.

Archaeologists have to be especially careful before, during, and after excavation because digging up a site totally destroys that site. Once you remove the dirt and artifacts, expose the features, and record what you find—that's it. You have demolished the

"I loved working outside and the process of excavation. It gave me a feeling of physical satisfaction to be tired at the end of the day. I also knew we were working toward a greater knowledge of the past. My least favorite element of archaeology is backfilling, which I think is self-explanatory."

—Perri Gerard-Little, 22, recent graduate of Columbia University provenience (place that artifacts are found), and you can never get it back. Knowing where an artifact was found in the ground is absolutely critical to understanding it. You must know how it relates to other cultural material and how deep it was located so you can determine how old it is.

If you've ever been to a natural history museum, you might have been impressed by the artifacts on display. Maybe you saw haunting jade masks, elaborately woven fabrics, colorfully painted pottery, or ancient gold coins. While these items are worth viewing, they are of almost no historical value unless their pedigree can be verified. The only way to know the context in which they were found is if their excavation was well-documented. The background of an artifact is what makes it significant.

For instance, finding a Euro coin in a Frenchman's pocket would be quite ordinary as that is their currency. Finding one in your home in the United States is far more interesting. Discovering a Euro in America suggests that you traveled long distances, engaged in world-wide commerce, and kept a memento of your vacation. If we were archaeologists from the future, we'd find that you had access to transportation—likely an airplane—had first-hand contact with another culture group, and had exchanged merchandise or information. In short, an artifact with no pedigree might be neat to look at, but it's not scientifically noteworthy. When you know where an artifact is from and when it was found, it becomes a clue in a story about human beings.

The stratigraphy (dirt layer) an artifact is found in will let an archaeologist know how old the artifact is. It is generally assumed that the deeper something is buried, the older it is. The reason is because dirt builds up over time. Even when you clear dust off of your computer screen or TV, it seems to return right away. The same process occurs outside. In time, things that were once on the surface of the ground get covered up by dirt and vegetation. Rain helps push artifacts down. Human practices also contribute to this process. For instance, new buildings are often constructed on top of old ones.

Exercise 2.9: Can you think of any exceptions to this rule? Where might the depth of a buried artifact not depend on when it was placed in the ground?

_____

_____

_____

Knowing how artifacts relate to one other in the dig is just as important as knowing their relative age. The place, location, and relationship of artifacts are critical to understanding their story. Think of the example of the Euro coin from earlier. A Euro found in Europe is not as significant as knowing it was found in America. You had to know *where* that Euro was found before it revealed information about its owner. The same thing goes for artifacts at Pambamarca. After all, we're dealing with three culture groups: the Cayambe, Inca, and Spanish. All of them interacted, but the question is how, where, and under what conditions? Provenience information for each artifact is therefore necessary.

Exercise 2.10: The Spanish brought animals, plants, and technology from the Old World that simply weren't available in the Americas. Name at least three of these things. You can consult the internet if you don't know any.

_____

_____

_____

Think about this scenario. The hacienda owners of yesteryear were notoriously cheap. They didn't feed their indentured servants. Instead, they gave them some land on which to grow their own food. We know that by the time the Spanish arrived, the Inca had come to Pambamarca. Therefore, indentured servants working for the Spanish could have been Inca or Cayambe. If the Spanish wouldn't even feed their servants, they certainly didn't share technology. The tools these people used to plow their fields and grow their crops were probably the same ones they were using well before the Spanish arrived. That said, finding a stone tool doesn't necessarily say anything about who used it or when. A stone tool could have been used during pre-Inca times, Inca times, or Spanish times. The only way to know the difference is to understand how it was associated with other artifacts.

Exercises 2.11: Look at Figures 2.17 and 2.18. They both show the stratigraphic layers in an excavation unit and the artifacts found in each layer. The artifacts are the same for both figures, but where they were found differs. For each figure, answer the questions that follow.

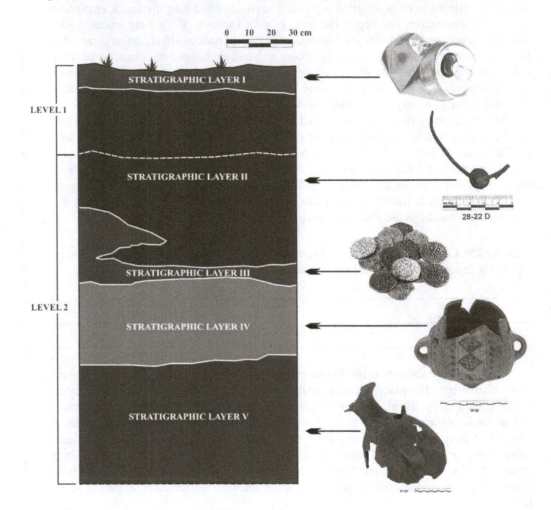

FIGURE 2.17

Who used the site? _____

How did land use change over time? _____

_____

Was this site used by only one group? Was there any overlap? How can you tell? _____

_____

_____

_____

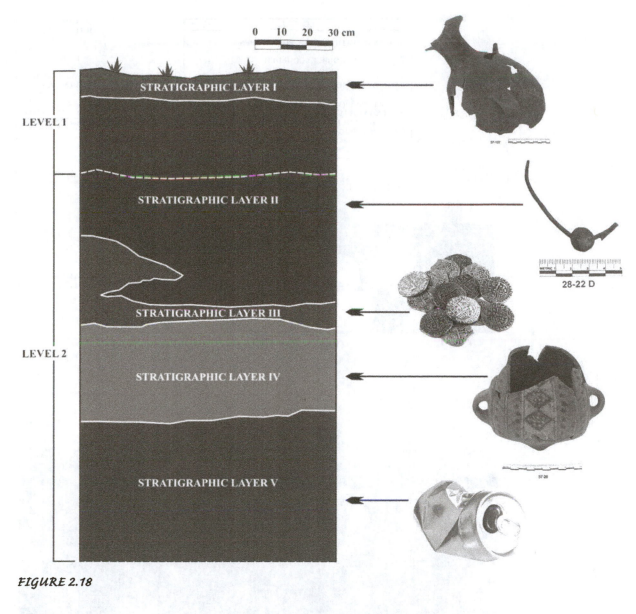

FIGURE 2.18

Who used the site? _____

How did land use change over time? _____

_____

Was this site used by only one group? Was there any overlap? How can you tell? _____

_____

_____

_____

Detailed recordings of where you dug, in what layers, and to what depth is, therefore, extremely important. You also have to take pictures and make profile drawings of the dirt layers (Figure 2.19). Doing this might seem time consuming or like detail overload, but it's what makes *interpretation* of a site possible. Yeah, it's fun to dig holes and get your hands dirty, but being a detective requires careful observation of clues. Remember, we want to do far more than collect old stuff. We want those artifacts to reveal prehistory. Without knowing crucial information about where and how the cultural material was found, we won't be able to piece history back together.

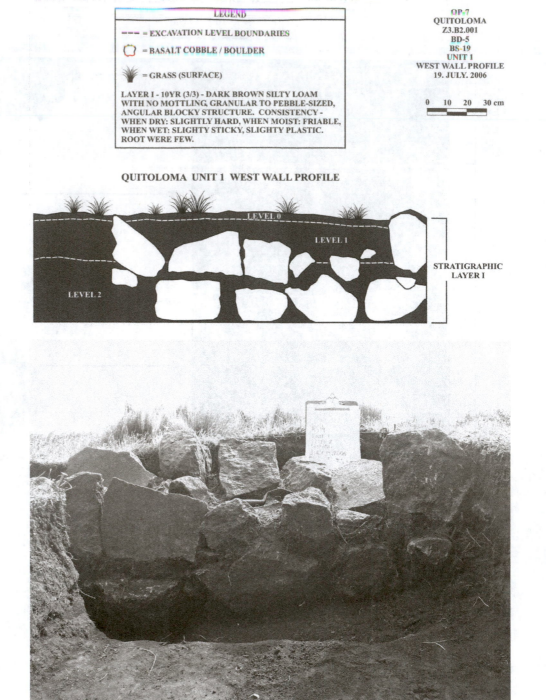

OP-7
QUITOLOMA
Z3.B2.001
BD-5
BS-19
UNIT 1
WEST WALL PROFILE
19. JULY. 2006

FIGURE 2.19   This is a sample of the types of data collected. The sketch is a scale drawing of what the photo shows. Note that detailed soil information is also included. Courtesy of PAP/Samuel Connell.

Another challenge of excavation is interpreting features as you dig. A feature could, for instance, be a wall, a trash pit, a post hole (hole where a post once held up a roof), a door, or a cache of weapons. Imagine you are digging an excavation unit. You get down 30 cm below the surface and you encounter a bunch of rocks in half the unit. Some rocks are big, some are small, some are sticking up above the others. Here's the question: have you hit a rock wall or building foundation of some sort? Or, are you just digging through naturally occurring stones in the area? Or was this part of a building that tumbled down a long time ago? Is it man-made or natural? Determining what you're seeing as you dig is a challenge.

Exercise 2.12: This next exercise asks you to determine what you're looking at in the excavation. Does it look natural or manmade? If it's manmade, what do you think it is? Explain your reasoning. If it's natural, explain why you think that. Respond to these questions after every figure (Figure 2.20–2.27).

FIGURE 2.20   Courtesy of PAP/Samuel Connell.

_____

_____

_____

FIGURE 2.21   Courtesy of PAP/Samuel Connell.

_____

_____

_____

FIGURE 2.22   Courtesy of PAP/Samuel Connell.

FIGURE 2.23   Courtesy of PAP/Samuel Connell.

FIGURE 2.24   Courtesy of PAP/Samuel Connell.

_____
_____
_____
_____
_____
_____

FIGURE 2.25   Courtesy of PAP/Samuel Connell.

_____
_____
_____
_____
_____
_____

FIGURE 2.26   Courtesy of PAP/Samuel Connell.

_____
_____
_____
_____
_____
_____

FIGURE 2.27   Courtesy of PAP/Samuel Connell.

_____
_____
_____
_____
_____
_____

FIGURE 2.28   Students work on laptops, consult maps, and catalog artifacts in a large space at the Hacienda Guachalá. This is the field lab. Courtesy of PAP/Samuel Connell.

STEP 3: LAB WORK

Lab work is where all the careful measurements, photographs, and recordings you made in the field come to life. This is where artifacts are sorted, analyzed, and tested. This is where the clues left by historical people start to "speak." You can think of this as the CSI headquarters for archaeologists. At Pambamarca, our lab is located in the Hacienda Guachalá (Figure 2.28). This is also where the artifacts are stored, so they are available to anyone in the community who wants to see them.

There is a basic procedure that all artifacts go through. When they first come into the lab, they are washed. As you can imagine, thousands of years under ground make them very dirty! They have to be washed so that paint on pottery or butcher marks on bones can be seen. After they are washed, they are sorted, counted, and tabulated. Figure 2.29 is an example of the information recorded. The distribution of artifacts through stratigraphic layers is significant. For instance, does any one layer contain more artifacts than another layer? If so, this might have been when the site was used the most. Perhaps it was only intermittently used in earlier or later times.

In addition to counting artifacts, they also have to be analyzed. Distribution patterns are useful, but they have their limitations. For instance, maybe we know that there is a significant amount of pottery in Layer 2, but what kind of pottery? Who made it? When was it used? Prehistoric groups weren't kind enough to sign their work so we'd know who made it. Instead, archaeologists study the types of paint, designs, temper (material used to keep clay glued together), and shapes of pottery. Based on this information, they know who made the pot. Understanding how paint, design, etc., relate to culture groups and time periods is called seriation.

Seriation patterns at Pambamarca are still being figured out. However, there are rules of thumb. For instance, the Spanish often glazed their pottery with a greenish or yellow color. The Inca made designs on their pots, such as little leaves or lines with balls at the end (Figure 2.30). The Inca also made large jars with pointy bottoms that had little

ID	Operation	Catalog #	Site #	Built Division	Built Space	Unit	Level	Feature	Locus	Descrip1	Descrip2	Descrip3	Count	Wgt (grams)	Date	Comment
1970	22	308	Z3.B2.007	3	0	15	0	0					1	10.10		listed as other, modern, trash
1954	22	292	Z3.B2.007	3	0	15	0	0		Bone			1	1.00	20-Jul-06	
1960	22	298	Z3.B2.007	3	0	15	1	0		Bone			1	0.50	10-Jul-06	burnt tool
1962	22	300	Z3.B2.007	3	0	15	1	0		Bone			18	6.20	10-Jul-06	burnt
1987	22	309	Z3.B2.007	3	0	15	1	0		Bone			22	46.40	18-Jul-06	
1974	22	312	Z3.B2.007	3	0	15	1	0		Bone			101	212.60	21-Jul-06	
1979	22	316	Z3.B2.007	3	0	15	1	0		Bone			6	29.40	21-Jul-06	
1983	22	320	Z3.B2.007	3	0	15	1	0		Bone			1	0.50	21-Jul-06	
1991	22	324	Z3.B2.007	3	0	15	1	0		Bone			1	0.50	21-Jul-06	
1538	23	148	Z3.B2.007	2	1	15	1	0	0	Bone	Burnt		4	5.00	01-Jul-05	
1952	22	290	Z3.B2.007	3	0	15	1	0	pit feature	Ceramic			14	927.00	20-Jul-06	
1955	22	293	Z3.B2.007	3	0	15	1	0		Ceramic			3	11.00	20-Jul-06	
1958	22	296	Z3.B2.007	3	0	15	1	0	pit feature	Ceramic			3	83.00	20-Jul-06	
1963	22	301	Z3.B2.007	3	0	15	1	0		Ceramic			9	22.00	10-Jul-06	
1968	22	306	Z3.B2.007	3	0	15	1	0		Ceramic			9	124.30	18-Jul-06	surface collection
1972	22	310	Z3.B2.007	3	0	15	1	0		Ceramic			139	1176.60	20-Jul-06	
1976	22	313	Z3.B2.007	3	0	15	1	0		Ceramic	Non-Diagnostic		3	3.00	21-Jul-06	
1977	22	314	Z3.B2.007	3	0	15	1	0		Ceramic	Diagnostic	Rim	1	316.00	26-Jul-06	1 burnt rim
1980	22	317	Z3.B2.007	3	0	15	1	0		Ceramic	Non-Diagnostic		27	858.50	21-Jul-06	20 burnt
1981	22	318	Z3.B2.007	3	0	15	2	0		Ceramic	Diagnostic	Rim	5	79.50	21-Jul-06	3 burnt
1982	22	319	Z3.B2.007	3	0	15	2	0		Ceramic	Diagnostic	Handle	1	11.50	21-Jul-06	broken off piece of handle
2005	22	325	Z3.B2.007	3	0	15	2	0		Ceramic			1020	8644.10		large sample -- 4 bags.
1995	22	330	Z3.B2.007	3	0	15	2	0		Ceramic	Historic	Glazed	4	4.89	10-Jul-06	
2003	22	338	Z3.B2.007	3	0	15	2	0		Ceramic	Diagnostic		1	9.50	13-Jul-06	
1978	22	315	Z3.B2.007	3	0	15	2	0		Groundstone	Igneous		1	823.00	21-Jul-06	

FIGURE 2.29   An example of the recordings made for each and every artifact as it is checked into the lab.

57-26

FIGURE 2.30   Example of Inca pottery showing typical design pattern. Courtesy of PAP/Samuel Connell.

FIGURE 2.31   Sam shows students large cone-bottom Inca pot. Courtesy of PAP/Samuel Connell.

camel-shaped nubs on the sides for tying on rope (Figure 2.31). The Cayambe didn't make jars like this, nor did they paint designs on their pottery. They might have added a red slip, but that's it. The Cayambe didn't have pottery wheels, so the construction of the pots is also something to consider. Knowing this type of information makes it possible to identify who used what pot, based on design.

Recently, some members of PAP decided to do an analysis of the mineral composition of obsidian stone tools at Pambamarca. If you know the mineral composition of the obsidian, you can compare it to the mineral composition of obsidian quarries to determine where the tools came from. Both the Inca and the Cayambe used obsidian to make cutting or scraping tools, though neither group used it to make weapons.[1] However, there was no way to tell who made the tools just by looking at them because both the Cayambe and the Inca made tools in the exact same way. Therefore, if you found obsidian in an excavation, it wouldn't give much information about the site. This new mineral composition analysis changes that.

If the archaeologists could figure out where the obsidian was coming from, they could learn a lot about Inca and Cayambe relationships. For instance, were the Inca carrying their obsidian up from Peru? Were they stealing Cayambe tools? Were they making their own from local sources? Knowing where the obsidian came from might explain when and how the Inca were able to conquer new territory. For instance, we know that the Inca were moving north from Cuzco into Ecuadorian territories. We might expect their obsidian sources to come from the south, but when did they start using locally sourced obsidian? If they had access to those quarries, it might imply when they ruled that territory.

Bone analysis is another crucial element of laboratory analysis. Osteology reports could be made for human burials. Physical anthropologists might be able to tell how the person died or if he/she suffered from any diseases, malnutrition, or stress in life by

"Lab work is a very important part of archaeology. We can excavate all we want, but if we don't conduct lab work, we will never learn anything. In the past we have lost artifacts simply by not cataloguing them correctly, and ultimately, the data is lost. The nature of a site is understood by the artifacts, stratigraphy, location, and architecture, amongst other details. I find lab work fascinating because I get to see firsthand what the people were doing, eating, making, and how they were living. This is why we do archaeology!"

—Ana Gonzalez, University of Hawaii, co-director of PAP

looking at the bones. If many burials are found with disease markings, that might indicate an epidemic swept through the area at some point in history. Animal bones are also of interest. Animal bones show what people ate or how they hunted, depending upon butcher marks or tools found associated with the kills. Were any new animals introduced to the area? Did people make bone tools?

These are examples of the types of questions that can be answered by studying artifacts. Learning as much as possible about what materials they were made from, where those materials are found, etc, gives critical information to archaeologists. Based on what you know from this chapter and Chapter 1, do Exercise 2.13.

INSTRUCTOR'S NOTE: This exercise is a fun small group activity. Students can strategize together about the artifacts, culture groups, and time periods. Make it a race: which group completed the exercise first? Groups can compare their answers. As a class, discuss why each group drew the conclusions it did. The goal of the exercise is to practice reasoning skills and to improve how that logic is articulated.

Exercise 2.13: Figure 2.32 shows different artifacts excavated from various sites at Pambamarca. Look carefully at each one, labeled A–G. First ask yourself what the item is and what it's made out of. Knowing the material might help you determine which culture group used it. Next look for any distinguishable marks or designs. You were given enough information in Chapters 1 and 2 to make educated guesses about the following questions. For each artifact, indicate what culture group used it and during what era (the eras can be pre-Inca times, Inca times, or Spanish times). Could any artifact be used by multiple groups? You will need to explain your reasoning for categorizing the artifacts.

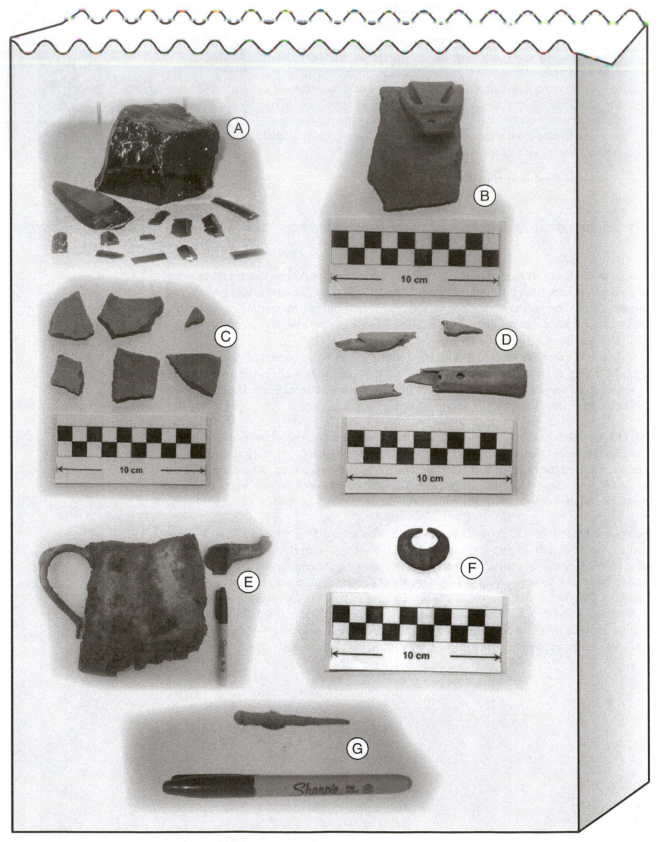

FIGURE 2.32   Bag of Artifacts.

Artifact A

What is the artifact? _____

Culture Group(s): _____

Time Period: _____

Reasoning:

_____

_____

_____

Artifact B

What is the artifact? _____

Culture Group(s): _____

Time Period: _____

Reasoning:

_____

_____

_____

Artifact C

What is the artifact? _____

Culture Group(s): _____

Time Period: _____

Reasoning:

_____

_____

_____

Artifact D

What is the artifact? _____

Culture Group(s): _____

Time Period: _____

Reasoning:

_____

_____

_____

Artifact E

What is the artifact? _____

Culture Group(s): _____

Time Period: _____

Reasoning:

_____

_____

_____

Artifact F

What is the artifact? _____

Culture Group(s): _____

Time Period: _____

Reasoning:

_____

_____

_____

Artifact G

What is the artifact? _____

Culture Group(s): _____

Time Period: _____

Reasoning:

_____

_____

_____

Chapter 3
Applied Anthropology and Ethics

This chapter is about applied anthropology and ethics. Applied anthropology is anthropology for the "real world." Sometimes this means doing anthropology for a specific client or purpose (outside of academia), but it also implies community involvement. An archaeologist might raise money or support requests for governmental aid to assist the town or village he/she works in. Although this type of activism is by no means mandatory, there is a growing trend to at least consider how archaeological fieldwork might impact the local community. In the past, anthropologists overlooked the desires of indigenous groups in order to press forward with "pure" science. Now days, it's more common to keep their interests in mind.

Of course, not all archaeologists feel an obligation to work for their community. Many come only to do fieldwork and think it's better to not stick their noses in other people's business. They come for science and limit their responsibilities to their work. On the other hand, PAP (among other projects) acknowledges that merely being present in off-the-beaten path places is an impact in and of itself. PAP feels that bringing a large sum of money and foreigners to a very isolated society will affect it. Therefore, we feel a duty to maximize the positive influence of our presence. Then again, even the definition of *positive* opens up a Pandora's box of ethical issues.

For instance, who decides what is positive and for whom something is "good"? What is the role of the archaeologist in the local community? Does the archaeologist have any responsibilities to the native population? What if they don't want you to dig somewhere? What if not digging would ruin your chances of answering your research questions? Would you ever sacrifice science to appease the community you work in? You'll confront the real-life challenges of juggling differing opinions, stakes, and interests at an archaeological site in this chapter. We'll explain the historic relationship between anthropologist and native person so you can draw educated conclusions about how archaeology should be conducted.

Finally, in order to understand how life is changing in the Pambamarca region since the arrival of PAP, we will discuss some local traditions such as Inti Raymi (also known as the Sun Festival, Summer Solstice, Fiestas de San Pedro and San Pablo), local food, music, and minga, which is a way of doing work. This will give you an important glimpse into the lives of the people in this area. Then we'll talk about how things are changing. We'll discuss who the major stakeholders in the Pambamarca Archaeological Project are, and you'll be given a chance to consider everyone's point of view. Maybe you'll even have a mock community meeting in your own class to debate the pros and cons of archeaology in Pambamarca.

BEHIND THE SCENES

Sometimes professional archaeologists don't like to admit it, but as children, lots of them were influenced by the movie series, *Indiana Jones*. (At least, one of the authors of your book falls into this category!) If you haven't seen *Indiana Jones* (maybe you can convince your professor to show you a clip), the films are about a swashbuckling archaeologist who travels the world in search of awe-inspiring ancient sites and artifacts. "Indy" will pop into Egypt and explore a forgotten pharaoh's tomb or crawl through a South American cave to recover pre-Columbian gold.

The movies are pure adventure, titillating adrenaline, and usually end with Dr. Jones being chased away by loin-cloth clad natives armed with spears or men in tunics with guns. Why is Indiana being chased? Because he stole the precious artifacts of local communities without permission. In fact, he had no permits to dig, and he doesn't appear to respect local communities, their opinions, or their history.

Indiana is not a very modern archaeologist, and unfortunately, even his good looks won't let him get away with professional misdeeds. He lands in a country, digs where he wants, and takes the loot. These days there are international laws, codes of ethics, and an anthropological responsibility to consider how your work as a scientist might impact the lives of the very real community in which you conduct research.

Oftentimes how an archaeologist deals with the community is considered such a behind-the-scenes issue for project directors that it's not discussed with field school students. To the authors of this book, that's like ignoring the 800 pound gorilla in the room. As a field school student, you will interact with the local community (Figures 3.1 and 3.2) and certainly witness how archaeologists get along with the people they work near. There is a long and sometimes notorious relationship between archaeologists and indigenous communities that drives modern decisions about these relationships.

FIGURE 3.1 Student dancing at fiestas in borrowed chaps. Courtesy of PAP/Samuel Connell.

FIGURE 3.2 Archaeology students paired up with their *"pequenas guias,"* or their little local guides for the summer. Courtesy of PAP/Samuel Connell.

A BRIEF HISTORY OF ARCHAEOLOGY IN THE NEW WORLD

Archaeology's history of imperialism is far more than postmodern PR mumbo jumbo. Think about the following words alone: *Indian, New World,* or *pre-Columbian*. From the moment Christopher Columbus and other Europeans set foot in the Americas, they have had the opportunity to name, define, label, and stereotype local people. The repercussions of these policies have echoed through centuries. Archaeologists were charged with writing the history of the indigenous Americans, and because most New World archaeologists were of European decent, Native Americans have been viewed through the lens of the colonist.

I interacted with the locals on a daily basis, as did most of the PAP members. I think this was important because we needed to build a relationship with these people. We did not want to seem like intruders to their homes. I thought it was a great complement to the trip because it allowed for a more complete experience.

—Danielle Babcock, 23, current employee of Concord Academy though PAP member two years ago as a senior at Mount Holyoke College

I interacted with many of the local children, and as a result, I became friends with members of the community. I even went camping with them. I feel the experience complemented my overall experience—in fact, it made it better. It was nice to get to know people and speak in my native tongue. PAP does so much for the people in the area, and this is an aspect that should be kept!

—Sorayda Santos, 23, UCSD Alumni, current Caregiver for the County of San Diego

Tensions between anthropologists and native groups started many years ago. In the late 18th and early 19th centuries, European-Americans wanted to learn more about Indian origins. Sounds innocent enough, right? Not so fast. During this time period, people were trying to understand the natural world by using scientific methodology. Amassing data about the Indians required describing, classifying, and amassing samples of . . . humans. If scientists expected to comprehend the Native American, they needed specimens, and the greatest collection of Indian specimens was found in graveyards. Naturally native groups protested scientists digging up their ancestors' remains, but that didn't stop "scientific research."

As a result of the anthropologists' craving for knowledge, hundreds of Indian burials were stolen for natural history museum stockpiles.[1] The culture, wishes, and religion of these people were not respected—or even acknowledged. Scientists were interested in native people, though not as human beings. Their value was in their physical remains. Studying the bodies of Indians was an opportunity to discover the biological nature of the "other." Scientists were trying to quantify exactly how different an Indian was from a European. Native Americans were viewed as a curiosity to be studied and documented. No more.

If grave robbing seems to be of dubious moral nature, then surely the next examples put anthropology on even more tenuous ground. Around the turn of the 20th century, massive ethnographic (cultural) studies were being conducted on native populations around the world. However, the logistics of traveling to these remote sites was cumbersome, and anthropologists found it more convenient to import their test subjects. One such example comes from 1897 when six Eskimos (also called *Inuit*) from northern Greenland were brought to New York City so they could be studied more easily by American anthropologists. The Eskimos were invited to come, and they agreed.

When the native Greenlanders arrived in America, they became instant celebrities because they were the most popular *exhibit* at the natural history museum. More than 30,000 people paid 25 cents each to view real, live "living fossils"[2]—people thought to be physically and culturally preserved in time. Euro-Americans thought it was definitely worth a quarter to see present-day cave men. Anthropologists of that day looked at living indigenous people as *data*—perfectly acceptable to be kept as displays at a museum.

Unfortunately, this story gets a little worse as it continues. Four of the six Eskimos died within a year of tuberculosis. Of the group brought from Greenland, one child accompanied his father. Unfortunately, when this child, Minik, was eight years old, his dad died of TB too. After Minik's father had passed, the anthropologists arranged for the boy to conduct his culture's typical funerary rights. The boy said that his dad should be wrapped in cloth, a funerary mask was to be placed over his face, and his body buried underneath a mound of stones. Then little Minik uttered the customary Eskimo funerary prayers at his father's final resting spot. The anthropologists hovered around this burial, notebooks and pencils in hand.[3]

Although the scientists initially seemed sensitive to the boy's loss, what the anthropologists didn't tell Minik was that the boy wasn't praying over his father's body. Actually, it was a fake corpse. Minik's father's actual body was being dissected, the brain preserved, and the bones stripped of their flesh so they could go on exhibit at the museum. Minik never knew the truth until he was 15 years old. He'd stayed in New York, as there was no reason for the orphan to return home. Imagine his horror when he innocently opened the newspaper, only to see an article about his father's remains at a natural history museum.

Naturally, he petitioned to get them back. "I felt as though I must die then and there. I . . . prayed and wept. I went straight to the director and implored him to let me bury my father. He would not. I swore that I would never rest until I had given my father burial," he was quoted as saying in the paper.[4] The news described his father's actual final resting place as follows. "An upstairs room—at the museum—is his father's last resting place. His coffin is a showcase; his shroud a piece of plate glass. No quiet of the graveyard is there . . . and when the sunlight fades they turn on the electrical lights so that Minik's father may not have even the pall of darkness to his naked bones."[5]

When the newspapers asked the anthropologists why they didn't let the boy bury the remains of his father, they reportedly explained that they guessed "closure" was more about the funerary process, not what was done with the actual bones. In fact, the anthropologists said that they had every right to the bones since no one else had claimed them. Remembering that Minik was only eight years old when his father died, the researcher said, "Well, Minik was just a little boy, and he did not ask for the body. If he had, he might have got it."[6] They never did relinquish Minik's father.

There are countless other examples of anthropologists claiming native property for the purposes of science. In 1911, "the most uncivilized, uncontaminated man in the world"[7] was discovered outside a mining town in northern California. He was the last known survivor of an Indian tribe that remained relatively untouched by white men. This last Indian was called Ishi, and when he was found, he was draped in a piece of canvas for clothes, wore a wooden plug in his nose, had deer sinew hanging in his ears as earrings, and had singed hair—a sign of mourning. His entire family had been killed by people who hunted Indians.[8]

After initially being locked in a jail cell for the insane—he was a "wild" Indian after all—he was taken to a museum in San Francisco and put on exhibit. Ishi literally resided among the artifacts and would give Sunday afternoon demonstrations on flint knapping, bow and arrow hunting, and tool making. From him, anthropologists learned all about his dead tribe, the Yahi, and recorded his language.[9] Because their subject was conveniently located in the museum 24/7, research was handy and readily available. Eventually the anthropologists proposed that Ishi stay with other Native Americans on a reservation. However, he chose to stay at the museum rather than live among strangers.

Instructor's Note: The following could also be small group discussion points.

Exercise 3.1: Would you let scientists dig up and study your dead relatives or friends? What are the circumstances that would change your mind one way or the other?

_____

_____

_____

_____

_____

_____

Exercise 3.2: Reality television is extremely popular these days. Often, the audience sees intimate details of people's lives displayed on camera and subsequently in gossip magazines and on websites. Do we still study humans in captivity? Explain. Is reality TV an extension of a zoo? Explain. How does it compare or contrast to Ishi's experience?

_____

_____

_____

_____

_____

the popularity of social networking sites like Facebook and Twitter, we can timate and even mundane aspects of our lives with others. Do you think tural tendency to want their lives documented? Explain why or why not. What o sharing?

_____

_____

_____

_____

_____

_____

_____

Anthropologists: The Silver Lining

Although the examples of Ishi and Minik are decisively unethical, that isn't the end of the story. Although anthropologists of old did not respect human rights as we conceive them today, their studies did help public understanding of native people. The prevailing ethic of the time was that Indians were either too savage or too simple to have built some of the major mound and temple complexes found in the New World. They were also considered to literally be walking prehistory: actual living examples of less-evolved man.

A prevalent hypothesis of the time was that certain human "races" were not as biologically developed as other human races. Scientists saw clear differences in the way people looked: different skin color, hair texture, cranial shape, etc. They assumed these physical differences indicated that people were subspecies of each other, which is the definition of race. It was even hypothesized that some human species should be named "*Homo ferus*" (wild man) or "*Homo monstrosus*" (monster man).[10] If you're guessing that *Homo monstrosus* probably wouldn't have had equal rights with *Homo sapiens*, you'd be correct. These racial classifications were a way to justify the inhumane treatment of "the other" (often, people of color).

However, anthropologists were able to challenge these notions. Because cultural anthropologists and archaeologists spent so much time excavating, recording, and detailing past and present Indian cultures, they were able to prove conclusively (with facts and data) that indeed living Indians were just as "evolved" as Europeans: they simply did things differently. In fact, their skin color or cranial shape did not biologically determine their technological advancements or intelligence. In short, there was no "*Homo ferus*," just different looking *Homo sapiens* with varying tools and skill sets.

Anthropologists were also able to prove that the Indians were the descendants of the people who built the ancient mounds and complex fortresses found in the New World. Europeans were particularly impressed with big buildings, and they initially doubted the Indians had anything to do with such splendid architecture. However, once ethnographic, archaeological, and biological tests were done, these facts were proved beyond a reasonable doubt. The bottom line is that no one would have believed the creativity and ingenuity of the Indians if it wasn't proven scientifically.

Because anthropologists had Minik's father's brain in a jar, they knew that a Polar Eskimo's brain was equal in size and complexity to that of a European's. Because they had Ishi in a museum exhibit, they were able to record what would have otherwise been a lost culture and language. Ishi and Minik's father are preserved in the annals of history and as such have contributed to our modern understanding of variation, immigration, and culture. Science does solve problems, correct misinformation, and falsify rumors. Studying mankind leads to a better understanding of humanity, which has the potential to benefit everyone.[11] Albeit, the data needed to draw these conclusions should have been collected in a more humane and respectful way.

Exercise 3.4: Do you think that ethics in science has improved over the years or are we making the same mistakes in more subtle ways? Explain and provide an example.

_____

_____

_____

_____

Exercise 3.5: Do you think there's ever a case where the benefits of science outweigh the risks? Can you think of an example where testing on humans, human embryos, or human remains is worth it? Why or why not?

_____

_____

_____

_____

_____

Exercise 3.6: Would you volunteer your body to be tested on? If so, what kinds of tests would be okay with you? What ones wouldn't be? If not, why not? Would you donate your body to science after you die? Why or why not?

_____

_____

_____

_____

_____

_____

Exercise 3.7: There have been major breakthroughs in ape language acquisition. Some chimps have been taught sign languages (ASL or others) to express limited wishes, feelings, and requests. We know there are limits to what extent we can study fellow humans. What about other sentient beings? If a creature has perception, cognition, and can express pleasure or pain, should it be treated differently? If it can literally ask to be set free, would that make a difference? Why or why not?

_____

_____

_____

_____

_____

_____

Native Groups Assert Themselves

In the early 20th century, Native Americans decided they'd had enough assimilating or being forced to disappear to make room for manifest destiny. Though their religion, languages, and traditions had been outlawed for years, they decided it was time for a cultural revolution. They petitioned the government for their rights, and some became cultural mediators who explained the importance of their Indian heritage to non-natives. Eventually, this ushered in the Indian Reorganization Act of 1934, which basically said that Native Americans could practice their cultures for tribal identity and sovereignty.[12]

During the Civil Rights Movement of the 1960s and 1970s, the American Indian Movement (AIM) was formed, and there was a backlash against anthropological fieldwork. This group disrupted archaeological sites, saying that it was wrong to disturb burial grounds. A famous civil rights manifesto written by AIM leader Vine Deloria claimed that anthropologists "infest the land of the free, and in the summertime, the home of the braves."[13] Deloria further accused anthropologists of reliance on data, ethnography, and excavations until they are "tolerably certain they represent ultimate truth."[14] He demanded to know why these studies represented a better "truth" about Native American life than what an Indian himself knew about his own past.

Furthermore, AIM was offended that anthropologists sat on their hands when, in 1954, Congress was voting to stop sending federal services to the reservations. Deloria warned that Indians were no longer going to be available for ethnographic and archaeological research if anthropologists were going to ignore the rights and interests of the society they wished to study. He demanded that anthropologists "get down from their thrones of authority and PURE [sic] research and begin helping Indian tribes instead of preying on them."[15] In short, Native Americans were tired of being lab mice, experiments, or data. If anthropologists wanted knowledge from the Indians, then they were going to have to show quid pro quo.

Exercise 3.8: Now you have heard all sides of the argument. How can anthropological studies potentially help or benefit native populations? What fears might contemporary native groups have about being studied? How could these studies hurt?

_____

_____

_____

_____

_____

_____

_____

INSTRUCTOR'S NOTE: Pretend an anthropologist wants to study college students. Let the class debate the pros and cons of allowing an ethnographer into their lives. This could be done as a class-wide discussion or as a paper.

PAP'S ROLE AT PAMBAMARCA

Most modern archaeologists know the history of the contentious relationship between anthropologist and native group. That knowledge has made a difference in the present day relationship between parties. For instance, some anthropologists did respond to AIM's complaints, and they started to do more for the Indian community. Also, Native Americans have won political battles, such as having rights to their burials through NAGPRA (the Native American Graves Protection and Repatriation Act), but in other cases, conflicts still arise.[16]

Is there a way to improve this relationship? Is there a way to ask and answer scientific questions while still respecting local groups? You know from Chapter 1 that the Pambamarca region has been fighting imperialism since the time of Inca invaders. Will the people of Pambamarca welcome outsiders now? To what extent is PAP expected to give back? Should they have to? Is it necessarily the role of the anthropologist to be a community advocate? What are the benefits or detractions of accepting this responsibility? The task of this book is to expose PAP's way of doing things (for better or worse), and leave it to you to discuss the pros and cons of the process. To do so, you'll need to be introduced to the town of Cangahua, PAP's headquarters.

Meet the Locals

The town of Cangahua (Figures 3.3 and 3.4) is the poorest community in the Pichinicha Province. According to a recent Ecuadorian National Census, a whopping 93 percent of Cangahua's population lives in poverty compared to over 60 percent of Ecuador's total population.[17] Cangahua has about 14,000 residents, though that number includes all of the surrounding communities who use Cangahua town as a meeting place. There are 48 official communities around Cangahua according to the census, though PAP deals with as many as 55. Of these, 95 percent of the population is indigenous, and five percent is mestizo (a mix of Spanish and indigenous). The majority of the indigenous live in the outlying comunities, while the mestizo largely live in town.

Most indigenous work as subsistence farmers (Figure 3.5), but others raise a surplus of potatoes, onions, or cereals to be sold in local markets (Figures 3.6 and 3.7). Electricity didn't come to the outlying communities until the year 2000, and it remains spotty. A good job in Cangahua is considered work in the local rose gardens, which export roses all around the world, including to the United States. Many of our Valentine's Day long-stem roses come from here. Unfortunately, long exposure to harmful pesticides and other chemicals can lead to health problems among employees, as well as contaminates in the local food and water supplies. A small portion of the other community members work in the milk or tourism industries.

The economy of this area is heavily dependant upon horticulture, and it always has been. Hence, some of the major features of Cangahua social life revolve around festivals that mark crucial farming cycles. For example, when springtime comes, the people of the Cangahua Parish get ready for the *cosecha*, or harvest, and the accompanying celebration of summer's arrival. To celebrate the return of the sun, they hold annual festivals called Fiestas

FIGURE 3.3   An aerial view of the town of Cangahua. Courtesy of PAP/Samuel Connell.

FIGURE 3.4   Cangahua's town square with view of the Catholic Church. Courtesy of PAP/Samuel Connell.

FIGURE 3.5 Women coming back from fields carrying children and crops. Courtesy of PAP/Samuel Connell.

FIGURES 3.6 AND 3.7 Fruit and grains sold in nearby market. Courtesy of PAP/Samuel Connell.

del Sol (Cayambe name), Inti Raymi (Inca name), or the Fiestas del San Pedro y San Pablo (Spanish [Catholic] name). These parties revolve around the June 21st solstice (the longest day of the year), but they go on for a couple of weeks—not just one day. These festivities have taken on many different meanings over the centuries, but above all, they tie the indigenous to their ancestors through music and dancing (Figures 3.8 and 3.9) that has been taking place for hundreds of years.

After all, whether sovereign, under Inca rule, or Spanish rule; whether Catholic or practitioner of pre-Hispanic religions; or whether in modern or prehistoric times, the people of Cangahua have been farmers. No matter what century you live in, the sun is a critical source of energy for agriculture. Therefore, the Sun Solstice festivals have always been an

FIGURE 3.8 Cangahua local David playing the *zampona,* an Andean panflute. Courtesy of David Morin.

FIGURE 3.9 Man dressed in fiesta costume. Note that this mask has two faces: one in the back as well. Courtesy of David Morin.

important celebration in this part of the world. Despite name changes, the dancing, costumes, music, food, and drink of the festivities have perpetuated for ages (Figure 3.10).

Early in the morning of the appointed day, people from every one of Cangahua's 55 outlying communities don colorful costumes and begin a non-stop dance that slowly processes down the mountainsides, coalescing in Cangahua's town square. The dancers move in time to the rhythms of drums and haunting melodies of flutes, as the crowd is energized with *chicha,* a locally made corn beer (Figure 3.11).

Finally, they *tomar la plaza,* which literally means they "take the plaza." Each community goes to its pre-assigned dance circle where the people dance around and round (Figure 3.12). Seeing thousands of people dancing, drinking, and celebrating as their ancestors did is a dramatic site. The combination of woodwind and percussion instruments, barbequed meat and booze, and blue jeans and hand-crafted masks creates a full-sensory experience.

Sporting events complement the excitement of the festival: specifically the local style of bullfighting. Adapting from the Spanish bullfighting tradition, Cangahueños find a suitably aggressive bull, tie a colorful bandana around his neck, and put him in an arena. Brave, young local men (no archaeology students) show both valor and athletic ability when they jump into the ring and attempt to retrieve the bandana (Figures 3.13 and 3.14). The first person to successfully complete this task wins the *colcha* (a beautifully elaborate blanket) and the admiration of pretty ladies from across town.

Syncretism is a concept in anthropology in which people blend new and old customs or take a dominant concept

FIGURE 3.10 Roasting Andean delicacy, *cuye,* or guinea pig. Courtesy of PAP/Samuel Connell.

FIGURE 3.11   Thirsty for chicha anyone? Courtesy of PAP/Samuel Connell.

FIGURE 3.12   Indigenous groups dancing in their circle in Cangahua's plaza. Courtesy of PAP/Samuel Connell.

FIGURES 3.13 AND 3.14   Bullfights are a fair event when the fighter doesn't carry a sword—just fast legs and inventive escapes! Courtesy of PAP/Samuel Connell.

and make it their own. Some examples of this come from the festivals and the bullfighting. The festivals of the sun solstice were anciently held by people who actually worshipped the sun, but now they blend in a celebration of Catholic saints. Bullfighting in Spain is theatrical and usually involves killing the bull. In Cangahua, it's a sometimes very amusing sport and the bull stays alive. In both instances, a combination of old and new traditions come together to make an event unique to Cangahua.

There are other examples of syncretism in Cangahua. For instance, many people in the area, especially the women, wear traditional dress. The woman in Figure 3.15 is wearing a typical outfit: brightly colored skirt, shawl, blouse, and bowler hat. You might not think of this is as typical Western fashion . . . but it is! The Spanish outlawed the tunics worn by the Inca and provided these peasants blouses and skirts to wear. The bowler hat was introduced in the 1920s by the British. Although these outfits are associated with indigenous communities in South America, they were actually European by design. However, they have been adopted and changed by the indigenous people.

"We paired up with our 'little guides,' which was great. The kids really took to us interacting with them (soccer or baseball games), and it was nice to have that after a long day in the field. Sometimes it was hard to forgo a shower or a nap to hang out with your *guia*, but it was all worth it in the end. I feel that working with locals complements the time at PAP—especially during the various festivals that were held in the town square."

—Amber Kling, 29, Ph.D. student at State University of New York at Buffalo

"The directors heavily emphasized the importance of a close interaction and association with the local community. I participated in the *minga* where we helped to clean up trash in the public square and cemetery. I also danced in the Inti Raymi festival, practiced my Spanish with my good friend, Franklin, and tried guinea pig for the first time. I definitely had a good time, but I also learned the importance of learning about the past with the people that live there."

—Matt Schauer, 27, Ph.D. student at University of Illinois at Chicago

A new brand of syncretism might be happening in Cangahua right now. Until PAP showed up, Cangahua was a proverbial one horse town. It was a quiet, off-the-beaten path place. The town's people practiced their festivals and customs for tradition, fun, and a way to let off steam after the work of a harvest. Now that the seeds of a tourism economy have been sown, these traditions are being co-opted as a part of a wider political and economic framework in Ecuador. They are a potential marketing tool to promote tourism to the region. It's a case of syncretism because one can't help but wonder how market value on centuries-old traditions might affect them.

Exercise 3.9: Name an example of syncretism in your own family. Sometimes people blend traditions during the holidays—especially with customs, food, or religion. Perhaps your celebration of holidays differs from that of your grandparents. Ask them or your parents what it was like growing up. Report the contrasts to your own experiences here.

_____

_____

_____

_____

_____

_____

FIGURE 3.15 Mother and daughter in classic Andean daily wear. Courtesy of David Morin.

PAP: Remembering History

As we said earlier, Indiana Jones is a fun movie, but the man is not a great example of an archaeologist. In real life, archaeology is not a game of "finders keepers." Any archaeologist working abroad has to have visas and permits from the national authority (like the government and a Ministry of Archaeology, or similar body) to dig. The same types of permits are required to work in the United States, though the process varies from state to state. Generally speaking, any artifact found (with the exclusion of human remains) belongs to the land owner, not to the archaeologist. Archaeologists don't get to keep what they excavate. They examine artifacts in their labs, but usually they have to return the property to the local government or land holder. Yes, even gold.

For archaeologists, the name of the game is data, learning, and discovery. If the archaeologist finding "treasure" were to plunder it, he/she would first be arrested for illegal trade in historic property. Then that person's professional reputation would be ruined, and he/she would likely never have the right to conduct a field project again. Also, that archaeologist would almost certainly lose his/her job. Because of the negative downside, and the fact that stealing valuable goods is antithetical to the field, archaeologists are rarely even tempted.

The country of Ecuador won't let you conduct an archaeology project unless at least one of the directors of the project has a Ph.D. You also need an Ecuadorian academic partner. Sam and Ana are a match made in research heaven: he wrote the dissertation and she's an Ecuadorian national. You also have to write detailed research proposals and submit a report on all of your findings at the end of every year. Ana says the government was happy to have archaeologists come to Ecuador and even asked them to work at other sites, though they only had eyes for Pambamarca. (Such a welcome doesn't happen everywhere.)

The permit Sam, Ana, and friends got from the national government gave PAP permission to dig wherever they wanted—including on private property. Does that mean it's time to pack up the trowel and hiking boots and head for the hills? Not so fast. Even though PAP technically has the right to dig anywhere in the Pambamarca region, that doesn't mean it would. From the beginning, the project directors felt that working in communion with the local people was critical. Common sense assumes that no one would appreciate a team of foreigners arriving unannounced to dig up their farmland or backyards. Furthermore, PAP wanted their project to be symbiotic: useful for archaeologist and locals alike.

I'll let Chad Gifford, anther director of PAP tell you how he feels about it.

"The federal permit allows us to work within Pambamarca quite freely. However, we have never pushed our way onto private lands owned by an individual or community without investing tremendous time, energy, and resources into developing meaningful relationships with the landholders in question. This is a conscious and deliberate ethic we have developed on our project: work for local communities and work with their consent . . . or don't work at all."

As we discussed in Chapter 1, indigenous communities in the Pambamarca region might have been neighbors, but they weren't a homogenous group. They were a loose confederation who would band together if outsiders came to their territory. Things aren't so different today. PAP wanted to do the right thing and get permission from local groups to dig. However, it is a complicated task. After all, there are 55 communities surrounding Cangahua. That means 55 different interests, intentions, and desires to fulfill. This level of maneuvering is not for the faint of heart.

Sam Connell lays it on the line.

"Working in Pambamarca has frustrated every archaeologist that has ever worked in the region—most told us we were crazy for attempting it. Colleagues of ours who have come to work with us for a field season have always left confused and bewildered by the difficulties. The cowboy archaeology attitude, which is very definitely based in the imperialism of science and righteousness to be here . . . that's not the way it works with them. The indigenous folks don't respect it at all. We don't lecture to them unless asked. We share coffee and wonder about the cosecha, the current crop harvest."

Exercise 3.10: What if the potential roles were reversed? Would you speak to and work with an anthropologist interested in doing an ethnography (cultural study)? Would you invite them into your home and your life so that they could understand you? Why would you want to or why would you not?

_____

_____

_____

_____

_____

Exercise 3.11: What would your stipulations or ground rules be before you decided to work with an anthropologist? Is there anything that would be off-limits for discussion or observation? What might you gain from the experience?

_____

_____

_____

_____

_____

Who Benefits from an Archaeology Project?

Conducting studies regardless of what the local community thinks is over—at least at Pambamarca. Sam, Ana, Chad, and the rest of the PAP leadership wanted to approach their project as community-minded archaeologists. They didn't want to force themselves upon the people; they wanted it to be a give and take relationship. Overall, this dream appears to be coming true. PAP is generally well-respected and welcomed in the area (Figure 3.16). The economic and other social benefits it brings to town every year are likely a driving force behind this acceptance.

When the lost gringo, Sam, and his wife, Ana, first wandered in town, innocently asking directions to Quitoloma on a mere Sunday afternoon drive, perhaps Don Carlos recognized an opportunity. When they revealed they were professional archaeologists, maybe the locals saw this as a chance to prove that the twice-conquered Cangahueños had glory days: indigenous fortresses that held off the mighty Inca. It could have been a chance to reclaim their honored history or find a job with the foreign archaeologists.

Opportunity was knocking for Sam, Ana, and the local community. There were three outstanding examples that the local people actively tried to recruit Sam and Ana to work in their area. First, of course, is Don Carlos who asked directly. He even suggested they get carbon dates to finally prove native Cangahueños built the fortresses. He had a 400 page history of the area he'd written on his own typewriter and just needed some irrefutable evidence to support what he had already concluded about the area.

The next example came from the Hacienda Guachalá. Now no longer the workshop it was in the 1500s, it's become a charming hotel just a few miles from Cangahua. Sam and Ana stayed there on their first night in town and talked to the owner, who no doubt realized an archaeology crew working nearby could mean guaranteed summertime guests. Furthermore, the Hacienda owners are very interested in local history and tourism. They even built a sundial (Figure 3.17) right on the equator (only a few miles from the Hacienda) and have a small museum and exhibit about Ecuadorian history and identity there.

The final example of PAP being welcomed to the area came during one of the earliest field seasons, in 2001. When Sam, Ana, and their tiny field crew hiked up Quitoloma for the first time that season, they found something curious. The last time they'd visited, Quitoloma's walls had all been in ruins. The stones had eventually fallen after hundreds of

FIGURE 3.16 Sam and Ana at the baptism of their Ecuadorian godchild in Cangahua's Catholic Church. After nearly a decade working in the region, enduring bonds between archaeologist and local grow. Courtesy of PAP/Samuel Connell.

FIGURE 3.17   Sundial installed at actual equator, only a few miles from the Hacienda Guachalá. Courtesy of PAP/Samuel Connell.

years in a harsh climate and strong wind. However, this time, some walls were neatly stacked—tidied up, if you will. The local community, the Chumillos Central, no longer allowed their animals to graze within the walls of the fortress. They had cleared their animals out of the way and tried to "help" with the archaeology. Sam and Ana considered this the proverbial welcome mat.

Exercise 3.12: Name an advantage and a disadvantage to having Sam, Ana, and the others there every summer. Why might people want or not want them there? (For instance, if a big development is slated to happen in your neighborhood, will that necessarily mean profits for you and your family?)

_____

_____

_____

_____

_____

_____

Isn't There a Metaphor About Good Intentions?

Some modern archaeologists want to help the local communities in which they work. Because anthropologists often are from Europe, America, or other developed countries, they have the resources and the influence to build roads, aid the development of tourist infrastructure, or even to help nominate archaeological sites as UNESCO World Heritage properties. This status definitely makes a place more desirable for tourists.

However, the question is: does the local community want this? And which members of the community would it benefit? Surely the business-owning mestizo group would be interested, but how would the indigenous feel? How might it change the economy? Now, we did say that most of the people living around Cangahua live in poverty. However, what we didn't emphasize is that many of these communities don't really need money to survive. Lots of them are subsistence farmers. They grow their own food without having to be on a cash economy. How would infusing a cash economy into this system disrupt it? Would it be harder to earn money to buy food than it would be to just cultivate their own? The economy is difficult everywhere—that doesn't mean growing potatoes is.

Some have accused community driven archaeology of being an example of paternalism. Paternalism is the feeling that like father, you know best. It's when anthropologists decide what local people need to improve their lives instead of letting them decide for themselves. Why should outsiders know what local people need more than they do? If a team of foreigners came to your hometown and told you that if only you did this and that, your life would be better, how would you feel? Even if they had a good point, do you think your community would respond positively or negatively?

Others think community oriented archaeology is an example of "post-colonial apologetics."[18] This means that anthropologists still feel guilty about what their forefathers did to native groups, so they're trying to make it up to local people by deciding what's best for them. Yes, scientists honestly do want to help, but is it really helping the locals or anthropologists' consciences? Some call activism "imperial liberalism,"[19] which implies that even if their hearts are in the right place, anthropologists make local communities dependent upon them for survival by getting so involved in local politics and life.

In the past, PAP has tried to take a hands-off approach to its community involvement. The project directors always asked permission to work, and they made every effort to explain to the local communities who they are, what they do, and what their intentions are (Figure 3.18 and 3.19). Sam, Ana, and friends think that transparency has been a key factor for building local support. They bring in experts so that the town can make informed decisions about its own development. Otherwise, PAP stuck to the business of archaeology, and this approach worked rather well.

OPTIONAL RESEARCH OPPORTUNITY

Can you determine how Cangahua has changed? Get on the web and try searching for Pambamarca, Cangahua, Inti Raymi, etc. What is going on right now to promote archaeology, tourism, or indigenous rights?

Does your town have an information center like this? Does anyone use it? Is there important cultural or historical information about your neighborhood that people should know?

Compare and contrast Cangahua's changing community with your own. How has your community changed in the past 5, 10, 20, or more years? Ask older folks in your neighborhood.

FIGURE 3.18 Community members at a town meeting about archaeology. Courtesy of PAP/Samuel Connell.

FIGURE 3.19 Locals in line for sheep stew after town meeting. Calling a meeting is a big affair. Many people have to walk for hours (or longer) from surrounding communities to reach Cangahua. Refreshments must be provided. Courtesy of PAP/Samuel Connell.

FIGURE 3.20 Local workers at the staircase gate at Quitoloma. Courtesy of PAP/Samuel Connell.

79

Sam, Ana, and friends assume that the approximately $50,000 PAP brings into the region on a yearly basis is a factor that has smoothed over relations. Keep in mind that those who have a good job in the rose factories make about $200 per month. This puts the $50,000 amount into perspective. As we discussed in the first chapter, archaeological sites are spread all over the landscape. Few, if any, are in Cangahua town. Therefore, a challenge to transparency is figuring out which of the 55 communities to ask about working at a given site. It's not often immediately clear which of the communities "owns" the site.

Offering the owners of an archaeological feature money to work on their property has proven to be a valuable tool for getting their consent to dig (Figure 3.20). PAP's incentive to the community is hiring some member's to help with excavations. Hiring a local workforce usually is a win-win situation. The archaeologists get extra hands, and the community makes money. Everything is fine until the inevitable pops up: what if you don't hire the right people? Or enough people? What if the site is half owned by one community and half owned by another? What if they don't get along? What do you do?

"PAP tries its best in a tough world. We hope always that our friends and family will be part of the community, yet the dynamics of small town life are a wild ride. Of course, we only come into town for a month or so a year."

The second year, excavation was shut down at another site because of a dispute between the different communities as to who the land belonged to. It impacted my experience a little, but also it is part of the whole process of doing archaeology in this region.

—Siobhan Boyd, 35, staff member, Gardiner Museum

Inevitable Escalation

Though PAP intially kept a relatively low profile regarding community activism, the longer PAP works in Cangahua, the more influential it becomes. Local schools are now using PAP's data to teach their children about the prehistory of the area. A local interest in

FIGURE 3.21 Ticket booth built to take tourists to see Quitoloma. Courtesy of PAP/Samuel Connell.

culture and history has spurred involvement from outside groups. A North American NGO called Putney came in and helped the Chumillos Alto community construct an information center, and the American Embassy funded the building of a ticket booth (Figure 3.21) at the base of Quitoloma. A graduate student in museum studies was recruited to help locals build a site museum dedicated to Quitoloma and the surrounding fortresses. All of these projects were undertaken to encourage tourism.

As a result, the municipal government agreed to finally fix the roads leading to Quitoloma (some parts of it were literally pre-Columbian). The communities are now petitioning the government to improve water and electricity access in the area. Did pressure from archaeologists quicken the pace of these improvements? Perhaps. The town of Cangahua is promoting its cultural history by starting a website (cangahua.com), which discusses the Inti Ryami festivals and other heritage. Some people in Cangahua want to build a cultural center in the town's square, across from the Catholic Church.

PAP leadership freely admits that things were rolling smoothly when they took a backseat approach to local politics. They treated people with respect, but pretty much minded their own business. Of course, there were minor issues from time to time. People occasionally asked to see the government permits before allowing access to their lands. Others were fine with excavation—but not during planting or growing season. A few communities had some internal political issues that outright prevented archaeologists from working on their territory. Overall, the relationship between archaeologist and townsperson was mostly harmonious.

However, in recent years, PAP has stepped up its community involvement. PAP field school curriculum now requires students to do a community service project through the Applied Anthropology program at Foothill College, the school where Sam teaches. The student government gives a small amount of money for a service learning project. The archaeology students then create a sustainable development project for the people in the Cangahua area. For instance, one year the Foothill students had playground equipment installed in the park. Another year, they donated money for school uniforms.

These ideas sound wonderful in theory, but in practice, helping is often easier said than done. PAP's deepening commitment to the community has also led it to be more intricately involved in local politics—which is sticky anywhere. Finally, if you say you want to make a difference, people will expect you to do so. Failing to deliver—even if that failure is inadvertent or unintentional—has a real life impact on the people relying on the aid.

I interacted as a translator for the other students and some staff. One of my closest friends on the project was an Ecuadorian student, which helped me gain more insights. However, I once overheard that some of the workers at the site were going to search the area for gold after we left. I worried whether I should say something or if I had heard correctly. In the end, we did a quick excavation to show them there was nothing else worth digging for.

—Erin Rodriguez, 21, junior at the University of Pittsburgh

I really enjoyed being able to interact with local people. I do think that it complemented my experience with PAP, though I often wished we had just a little more free time for hanging out with the local folk.

—Laurie Bramlage, 27, Purdue graduate and graphic designer

Many people agree with our research, but we still have a handful who do not agree with what we do and don't want us in town. There are some that believe we are looters. Others think we are trying to take "treasures" or "gold" back home with us.

—Ana Gonzalez, University of Hawaii, co-director of PAP

For example, one aforementioned Foothill service learning project provided funding to purchase school uniforms for the local pupils. This was a great solution to a community problem because purchasing the required uniforms was an economic hardship on families. Unfortunately, the Foothill students inadvertently ordered uniforms from an unreliable company that didn't send the merchandise. It wasn't their fault, but it impacted the success of the project because local children who were counting on a uniform didn't get one. If the outcomes of projects aren't successful, PAP risks appearing to cry wolf. Good intentions have to straddle practical considerations, and the well meaning must deliver.

Another example of complications occurred when PAP got involved with the community cultural center slated to be opened in Cangahua's town square. PAP's involvement was met with disapproval. PAP was accused of promising the world and then failing to deliver. Furthermore, who was going to pay for this structure? The church had already promised resources, including land, but how much were the archaeologists going to contribute? And who was in charge of the vision, building, and financing of this goal? Who had the right to write history, and who would profit from it?

Because PAP doesn't want to make broad decisions for all of the stakeholders, sometimes it's very difficult to get anything done. After all, there are 55 indigenous communities to please as well as the mestizo townspeople, the hacienda owners, the government of Ecuador, academic standards, and field school students. With so many people to satisfy, micromanaging all of it would not be effective (and it'd drive them crazy to try). On the other hand, it's impractical to take a backseat on the management and distribution of crucial finances. There is a constant learning curve for the PAP directors. As their presence in the Pambamarca region has continued through the years, they've discovered a lot about how to deal with everyone involved, although there's still plenty to learn.

"Over time, we have cycled in and out of communities as they wrestle with the idea of having us work there. During this time we have learned that it is important to attend as many community meetings as possible to discuss our work. We must never assume the community is a monolithic entity."

The American Anthropological Association code of ethics says, "Anthropological researchers have primary ethical obligations to the people, species, and material they study and to the people with whom they work. These obligations can supersede the goal of seeking new knowledge."[20] Striking the right balance can be difficult. There are many stakeholder interests to consider. The Pambamarca Archaeology Project considers it paramount to do good work and to do that work while considering its impact on the local community. Sam, Ana, Chad, and the others would rather risk making mistakes than ignore their emphasis on community minded archaeology. So far, their students agree.

The final activity asks you to consider the needs of the various groups at PAP and to try to negotiate them. Good luck!

I thought living within the village was extremely helpful. Without learning about the present, learning about the past in the same area would be missing something.

—Perri Gerard-Little, 22, recent graduate of Columbia University

I think it was very important to remember the project's responsibility to the communities working with, and affected by, the archaeology. As outsiders, we are guests in the country and should not be doing archaeology for selfish reasons, but for a greater understanding that many people can benefit from—especially those affected locally.

—Hannah Sistrunk, 23, UC Berkeley grad, field tech for Far Western Anthropological Research Group

INSTRUCTOR'S NOTE: The following exercise can be done individually. If it's done individually, each student can write from the point of view of one or all of the stakeholders and turn in the exercise as a paper. This exercise can also be done in a small group, where each student in a group is a stakeholder and represents his/her issues to the group. Or finally, this can be conducted as a class-wide project. Small groups of stakeholders can meet and brainstorm about their interests. Then, the whole class can be set up like a forum where each stakeholder group presents their points of view and a discussion ensues. Perhaps all members of the class can then vote on how to best deal with the given issues. No matter how the exercise is directed, the space below is given as a place for brainstorming.

Exercise 3.13: There are many stakeholders with varying interests at PAP. The following are the main ones. Using the information presented in this chapter (as well as data from any other chapter or outside sources, if you're so inclined), put yourself in the shoes of each stakeholder. What are your primary objectives? What are you afraid of losing out on? Will someone getting their way impede your chances of getting what you want? What might you be willing to sacrifice or bargain to broker the best deal for yourself or your community? What other groups might you partner with? Use these questions as a platform to really think about who these stakeholders are and what they want. Use the information from the chapter to guide your opinions, but then go beyond the chapter. Imagine what people would say or think or want.

Stakeholder 1: The Archaeologist

The archaeologist is an academic. If he/she gets research grants to do science, he/she has to come up with answers. Who might the archaeologist have to answer to? What will get in the way of a productive research session? What considerations other than scholarly decisions might he/she have to make? What are the archaeologist's rights and responsibilities?

_____
_____
_____
_____
_____
_____
_____
_____
_____
_____
_____
_____

Stakeholder 2: The Mestizo Townspeople

The Mestizo townspeople live in Cangahua town. They tend to be considered higher class than the purely indigenous people. Mestizos are often town leaders or business people (e.g., cabbies, shop owners, potential tour guides, etc.). What is at stake for them if the archaeologists are there? What are some benefits? What are some downsides? Is it only the archaeologists with whom they might have positive or negative relationships? What about the other stakeholders?

_____
_____
_____
_____
_____
_____
_____
_____
_____
_____
_____
_____

Stakeholder 3: The Indigenous Communities

There are many (more than 50) indigenous communities surrounding Cangahua, and they typically don't have much political power. What do they want? What are their interests? What might persuade some communities to work with the archaeologists? What might prevent them from wanting to cooperate? What would they want out of the deal? Other than the archaeologists, who else might they have to work with? How would this affect their cooperation?

_____

_____

_____

_____

_____

_____

_____

_____

_____

_____

_____

_____

Stakeholder 4: The Hacendado.

The hacendado, or the hacienda owner, represents the top tier of Ecuador's class-based society. You learned about the hacendado in the first chapter. They are of Spanish decent, they are the land owners, and they represent particular interests. How do their interests mix with the other stakeholders' interests? To whom do they owe anything? What do they want?

_____

_____

_____

_____

_____

_____

_____

_____

_____

_____

Stakeholder 5: The Student

What do you want in an archaeology project? What would your primary goals be? Is it science? The community? To what extent does this balance aid or potentially detract from your experience? All students in the class should consider this stakeholder!

_____

_____

_____

_____

_____

_____

_____

_____

_____

_____

_____

 **INSTRUCTOR'S NOTE:** If students did outside research, this could be turned into a paper or group project. You could potentially give them the following scenario. The local mayor wants to excavate a large fortress site and develop it for the town. He has $100,000 of UNESCO money to back him. Here's the catch: the mayor is the hacienda owner.

CHAPTER 1 ENDNOTES

1. Murra (1946); Puento (1974:22); (Cobo 1979[1653]); Hyslop (1990)
2. Athens and Osborn (1974:1); Athens (2003:20)
3. Gifford (2003:1–2)
4. Wolf (1982); Haynes and Prakash (1991); Irschick (1994); Miller (1995); Thomas (1997)
5. Stoler (1989); Adas (1992); Bhadha (1994); Lawson (1994); Cooper and Stoler (1997); Gifford et al. (2002:3)
6. Athens (2003:5)
7. Clapperton (1993) [cited in Gonzalez et al. 2006:25]
8. Athens and Osborn (1974); Salomon (1986:124); Athens (2003:7–10)
9. Paz Ponce de Leon (1897[1582]:111)
10. Athens (1992:210)
11. Athens (1992:212)
12. Puento (1974:22); Hyslop (1990)
13. Murra (1946); Cobo (1979[1653])
14. D'Altroy (2003:45–47)
15. D'Altroy (2003)
16. D'Altroy (2003:208)
17. Hemming (1970:171)
18. D'Altroy (2003:210)
19. Cobo (1979[1653]:157–8)
20. Cobo (1979[1653]:158)
21. Hemming (1970:124–5)
22. Hemming (1970)
23. Newson (1995)
24. Hemming (1970:47)
25. Hemming (1970)
26. cited in Hemming (1970:363)
27. Keith (1971); Powers (1991)
28. Bonifaz (1995)
29. Bonifaz (1995)
30. Bonifaz (1995)

CHAPTER 2 ENDNOTES

1. Ogburn, Connell, and Gifford (2009)

CHAPTER 3 ENDNOTES

1. Thomas (2000:4)
2. Thomas (2000:78)
3. Thomas (2000)
4. New York World (1907) [cited in Thomas 2000:82]
5. Harper (1986)
6. Thomas (2000:82)
7. Ishi: The Last Yahi (1994)
8. Thomas (2000)
9. Thomas (2000)
10. AAA (2007)
11. Landau and Steele (1983)
12. Thomas (2000)
13. Thomas (2000:200)
14. Thomas (2000:200)

15. Thomas (2000:200)
16. See examples, Thomas (2000); Apgar (2004); Athens (2004); Fine-Dare (2005); Viotti (2005 a, b)
17. www.paho.org (Pan American Health Organization) and the http://www.inec.gov.ec/web/guest/inicio (Instituto Nacional de Estadistica y Censos)
18. Lowenthal (1988)
19. Shweder (2003)
20. AAA (1998)

REFERENCES CITED

Adas, Michael. 1992. From Avoidance to Confrontation: Peasant Protest in Precolonial and Colonial South Asia. In *Colonialism and Culture*, edited by N.B. Dirks, pp. 89–134. Ann Arbor: The University of Michigan Press.

American Anthropological Association. 1998. AAA Code of Ethics. http://www.aaanet.org/committees/ethics/ethcode.htm. Accessed January 2012.

———. 2007. RACE Project. http://www.understandingrace.org/home.html. Accessed September 2009.

Apgar, Sally. 2004. Group Has History of Protecting Graves. *Honolulu Starbulletin* (8/12).

Athens, J. Stephen. 1992. Ethnicity and Adaptation: The Late Period-Cara Occupation in Northern Highland Ecuador. In *Resources, Power, and Interregional Interaction*, edited by Edward M. Schortman and Patricia A. Urban, pp. 193–219. New York: Plenum Press.

———. 2003. Inventory of Earthen Mound Sites, Northern Highland Ecuador. Report on file at International Archaeological Research Institute, Inc, Honolulu, Hawaii.

———. 2004. "Setting the Record Straight" for Cachola-Abad and Ayau. *Hawaiian Archaeology* (9):119–122.

Athens, J. Stephen, and Alan J. Osborn. 1974. Prehistoric Earth Mounds in the Highlands of Ecuador: A Preliminary Report. On file at Instituto Otavaleno de Antropologia, Otavalo, Ecuador.

Bhadha, Homi. 1994. *The Location of Culture*. London: Routledge.

Bonifaz, Diego. 1995. *Guachala: Historia de una Hacienda en Cayambe*. Cayambe, Ecuador: Tecnioffset C. Imprenta.

Clapperton, C. 1993. *Quaternary Geology and Geomorphology of South America*. Amsterdam: Elsevier.

Cobo, Father Bernabe. 1979[1653]. *History of the Inca Empire: An Account of the Indians' Customs and their Origin Together with a Treatise on Inca Legends, History, and Social Institutions*. Translated and edited by Roland Hamilton. Austin: University of Texas Press.

Cooper, Frederick, and Ann Laura Stoler, eds. 1997. *Tension of Empire*. Berkeley: University of California Press.

D'Altroy, Terraence N. 2003. *The Incas*. Malden, MA: Blackwell Publishing.

Fine-Dare, Kathleen S. 2005. Anthropological Suspicion, Public Interest, and NAGPRA. *Journal of Social Archaeology* 5(2):171–192.

Gifford, Chad H. 2003. Resisting Inca Imperialism in Ecuador: The Archaeology of a Militarized Frontier. Grant proposal for the National Science Foundation.

Gifford, Chad, Samuel Connell, Ana Lucia Gonzalez, and Maureen Carpenter. 2002. Difficult Encounters in Pambamarca, Ecuador. Paper presented at the 21st Annual Northeast Conference on Andean Archaeology and Ethnohistory, University of Pittsburg, November 2–3.

Gonzalez, Ana Lucia, Samuel V. Connell, Chad Gifford, Rudy Larios, Brandon Lewis, Oliver Wigmore, Steven Williams, Virginia Popper. 2006. Proyecto Arqueologico Pambamarca Informe Preliminar de la Temporada 2005. On file at the Instituto Nacional de Patrimonio Cultural del Ecuador, Quito, Ecuador.

Harper, Kenn. 1986. *Give Me My Father's Body: The Life of Minik, the New York Eskimo.* Frobisher Bay NWT: Blackhead Books.

Haynes, Douglas, and Gyan Prakash. 1991. Introduction: The Entaglement of Power and Resistance. In *Contesting Power: Resistance and Everyday Social Relations in South Asia,* edited by D. Haynes and G. Prakash, pp. 1–22. Berkeley: University of California Press.

Hemming, John. 1970. *The Conquest of the Incas.* Harcourt Brace and Company, San Diego.

Hyslop, John. 1990. *Inca Settlement Planning.* Austin: University of Texas Press.

Irschick, Eugene F. 1994. *Dialogue and History: Constructing South India, 1795–1895.* Berekely: University of California Press.

Ishi: The Last Yahi. 2002. Sanachie Entertainment Group, 57 min. Documentary.

Landau, Patricia M., and D.Gentry Steele. 1983. Why Anthropologists Study Human Remains. Paper presented at the 41st Roger Byrd Lakota Sioux Plains Conference, Rapid City, Iowa, November 4.

Lawson, Michael L. 1994. *Dammed Indians: The Pick-Sloan Plan and the Missouri River Sioux, 1994–1980.* Norman: University of Oklahoma Press.

Lowenthal, D. 1988. *The Past is a Foreign Country.* Cambridge: Cambridge University Press.

Miller, Daniel. 1995. Consumption as the Vanguard of History: A Polemic by Way of an Introduction. In *Acknowledging Consumption: A Review of New Studies,* edited by Daniel Miller, pp. 1–57. London: Routedge.

Murra, J. 1946. The Historic Tribes of Ecuador. In *Handbook of South American Indians,* edited by J. Steward, pp. 785–821. Bureau of American Ethnology, Bulletin 143, United States Government Printing Office, Washington D.C.

Newson, Linda A. 1995. *Life and Death in Early Colonial Ecuador.* University of Oklahoma Press.

Ogburn, Dennis, Samuel Connell, and Chad Gifford. Provisioning of the Inca Army in Wartime: Obsidian Procurement in Pambamarca, Ecuador. *Journal of Archaeological Science* (36):740–751.

Paz Ponce de Leon, Sancho. 1897[1582]. Relacion y Descripcion de los Pueblos del Partido de Otavalo. In Relaciones Geograficas de Indias (tomo III), edited by Jimenez de la Espada, pp. 105–120. Tipografia de los Hijos de M.C. Hernandez, Madrid.

Puento, Geronimo. 1974. Probanza de Don Hieronimo Puento, Cavique Principal del Pueblo de Cayambe, de Servicios. Documentos para la Historia Militar, pp. 11–50, Direccion de Historia y Geografia Militar del E.M.C. de las FF.AA., Casa de la Cultura Ecuatoriana, Quito.

Salomon, F. 1986. *Native Lords of the Quito in the Age of the Incas.* New York: Cambridge University Press.

Shweder, R.A. 2003. *Why Do Men Barbeque? Recipes for Cultural Psychology.* Cambridge: Harvard University Press.

Stoler, Ann Laura. 1989. Rethinking Colonial Categories: European Communities and the Boundaries of Rule. *Comparative Studies of Society and History* 31(1):134–161.

Thomas, David Hurst. 2000. *Skull Wars: Kennewick Man, Archaeology, and the Battle for Native American Identity.* New York: Basic Books.

Thomas, Nicholas. 1997. Partial Texts. In *On Oceania: Visions, Artifacts, Histories,* edited by Nicholas Thomas, pp. 23–49. Durham, NC: Duke University Press.

Viotti, Vicki. 2005a. "State Probes 'Injury' to Iwi." Honolulu Advertiser, B:1, March 9.

———. 2005b. "Artifact Negotiation Urged." Honolulu Advertiser, B:3, March 16.

Wolf, Eric R. 1982. *Europe and the People without History.* Los Angeles: University of California.

Wolf, E., and S. Mintz. 1957. Haciendas and Plantations in Middle America and the Antilles. *Social and Economic Studies* 6:380–412.